RAPID REVISION NOTES

A LEVEL

INORGANIC CHEMISTRY

BY E.J. PERKINS B.Sc.

General Editor – Professor A.J.B: Robertson
Professor of Chemistry, King's College,
London

CELTIC
REVISION AIDS

Celtic Revision Aids
30–32, Gray's Inn Road,
London, W.C.I.X. 8JL

© C.E.S.

First Published 1982

ISBN 0 86305 115 4

Printed and bound in Great Britain by
Cox & Wyman Ltd, Reading

NOTE ON NOMENCLATURE

The recommendations of the main Examination Boards and of the Association for Science Education have been taken into account in the system of units used throughout this book. The A.S.E. publications 'Chemical Nomenclature, Symbols and Terminology' and 'S.I. Units, Signs, Symbols and Abbreviations' have been referred to extensively in the writing of the text.

In some sections of this book, M is used to indicate a metal atom, A is used to denote an anion, and X is used to denote a halogen atom.

GENERAL EDITOR'S FOREWORD

The Rapid Revision 'A' Level Series is designed for students preparing for G.C.E. 'A' Level, Scottish Highers, Intermediate University and similar examinations. They are comprehensive and may be used either on their own for revision or as a complement to set books and text books. The notes are organised to help students remember facts and to pin-point areas of difficulty. Practice questions are included at the end of each section with answers to the numerical problems at the end of the text. Where necessary, clearly worked through examples have been included in the text.

I am sure students will find these books very helpful.

A J B Robertson

M.A., Ph.D., D.Sc., C. Chem., F.R.S.C.

Professor of Chemistry,
King's College London,
University of London

Formerly Fellow of St John's College
Cambridge

AUTHOR'S FOREWORD

Rapid Revision Inorganic Chemistry covers the elements in detail and the summaries of the properties of the groups of the Period Table are particularly helpful. Throughout, the material is arranged so that the student can revise speedily and thoroughly. The book is especially helpful as an *aide memoire* throughout a sixth form course or for students working on their own.

There are two other books in this series which are invaluable for studying the other sections of the 'A' Level chemistry syllabus. These are *Rapid Revision Organic Chemistry* and *Rapid Revision Physical Chemistry*.

E.J.P.

Contents

1 ATOMIC STRUCTURE

Atomic theory

1 Matter is made up of small particles called **atoms**.

2 The atoms of the same element have the same **atomic number**.

3 Chemical reactions takes place between simple whole numbers of atoms.

An atom is the smallest part of an element which can take part in a chemical change.

A molecule is the smallest part of an element or compound which can exist independently.

Structure of the atom

An atom consists of a central **nucleus** containing **protons** and **neutrons** with **electrons** surrounding the nucleus. The number of electrons equals the number of protons so that the atom is electrically neutral.

	Approximate relative mass	Approximate relative charge
Electron	1/1840	-1
Proton	1	$+1$
Neutron	1	0

Atomic number is the number of protons in the nucleus of the atom.

Relative atomic mass of an element is the mass of one atom compared with the mass of one atom of $^{12}_{6}C$ which is arbitrarily given a value of 12·000. Most values are not whole integers because naturally occurring elements consist of isotopes.

Isotopes are atoms of the same element having the same atomic number, but different relative atomic masses because of the presence of different numbers of neutrons, e.g. hydrogen.

Isotope	Number of protons	Number of electrons	Number of neutrons	Atomic number	Relative atomic mass
Protonium $^{1}_{1}H$	1	1	0	1	1
Deuterium $^{2}_{1}H$ or D	1	1	1	1	2
Tritium $^{3}_{1}H$ or T	1	1	2	1	3

Arrangement of electrons

Electrons occur in **shells** around the nucleus of an atom and are distributed between different sub-levels or **orbitals**. The emission spectrum of an excited atom is not continuous. Discrete amounts of energy are emitted or absorbed (quanta) as electrons move from one energy level to another.

$$\text{Quantum} = \Delta E = E_2 - E_1 = h\nu = \frac{hc}{\lambda}$$

where h = Planck's constant = $6 \cdot 625 \times 10^{-34}$ J s, ν = frequency of radiation, λ = wavelength of radiation and c = speed of light (3×10^8 m s^{-1}).

Electrons have **wave properties**. An orbital is an **electron cloud** representing the probability of finding an electron at any particular position at any instant. An electron is characterized by **four quantum numbers:**

(i)   **n** denotes the **number of the shell**, the number of orbitals in a shell being n^2 with n being 1, 2, 3, . . . ;

(ii)   **l** denotes the **eccentricity of the orbital** (s, p, d or f);

(iii)   **m**, the **magnetic quantum number** takes account of the behaviour of the electron in an external field and has the values +1 to −1 including 0;

(iv)   **s** denotes the **spin quantum number** and is $\pm\frac{1}{2}$.

Pauli's exclusion principle: no two electrons can have exactly the same set of quantum numbers.

Types of orbitals

(i)   **K shell** (n=1) contains one spherical orbital (1s) containing up to two electrons ($1s^2$).

(ii)   **L shell** (n=2) contains one spherical orbital (2s) and, at a slightly higher energy level, three dumb-bell shaped orbitals $2p_x$, $2p_y$ and $2p_z$. The s- and p-orbitals contain up to 8 electrons ($2s^2$, $2p_x^2$, $2p_y^2$, $2p_z^2$).

(iii)   **M shell** (n=3) contains up to eighteen electrons (one 3s-orbital, three 3p-orbitals and five 3d-orbitals – $3s^2$, $3p^6$, $3d^{10}$).

(iv)   **N shell** (n=4) contains up to thirty two electrons (one 4s-orbital, three 4p-orbitals, five 4d-orbitals and seven 4f-orbitals – $4s^2$, $4p^6$, $4d^{10}$, $4f^{14}$).

Aufbau principle: electrons first occupy the lowest energy orbitals available to them, they then enter higher energy orbitals only when the lower energy orbitals are filled. The sequence of the filling is 1s 2s 2p 3s 3p 4s 3d 4p 5s 4d 5p 6s 4f 5d 6p 7s 5f 6d 7p.

Electrons are affected by different nuclear charges and different electronic charges so that energy levels which are close together (e.g. 5f and 6d) sometimes alternate in energy and therefore in order of filling.

Hund's principle: orbitals of equal energy are each occupied by a single electron before the second electron of opposite spin (m) enters the orbital. This means that electrons stay as far apart as possible.

Consider nitrogen, atomic number Z. The electronic configuration is $1s^2\ 2s^2\ 2p^3$ and this will be $1s^2\ 2s^2\ 2p_x^1\ 2p_y^1\ 2p_z^1$.

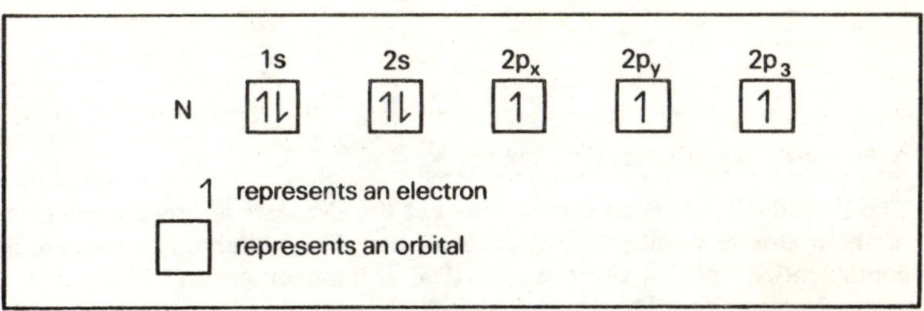

Paramagnetism is the property of attraction to a magnetic field shown by substances containing unpaired electrons.

Diamagnetism is the property of repulsion by a magnetic field and shows the absence of unpaired electrons.

Practice questions

1 Define an atom and a molecule.

2 Define atomic number and relative atomic mass.

3 What is an isotope? Illustrate your answer with references to hydrogen.

4 With reference to the arrangement of electrons in an atom, give the meaning of shell and orbital.

5 What is meant by 'quantized'? Give the four quantum numbers which characterize an electron.

6 State (a) Pauli's exclusion principle; (b) Aufbau principle; (c) Hund's principle.

7 Work out the electronic configuration of zirconium, atomic number 40.

8 From the following list of atomic symbols, choose (a) isotopes of the same element; (b) atoms having the same mass; (c) atoms with the same number of neutrons:

(i) $^{12}_{7}N$, (ii) $^{13}_{5}B$, (iii) $^{13}_{7}N$, (iv) $^{14}_{6}C$, (v) $^{14}_{7}N$, (vi) $^{15}_{7}N$, (vii) $^{16}_{7}N$, (viii) $^{16}_{8}O$, (ix) $^{17}_{7}N$, (x) $^{17}_{9}F$, (xi) $^{18}_{10}Ne$.

 9   Write the electronic configuration for $^{64}_{29}Cu$ taking into account that the configuration d^{10} is very favourable. Give also the electronic structures of Cu^+ and Cu^{2+}. State whether each is paramagnetic or diamagnetic.

10   Chlorine consists of two isotopes $^{35}_{17}Cl$ and $^{37}_{17}Cl$ and has a relative atomic mass of 35·5. Calculate the ratio of atoms of $^{35}_{17}Cl:^{37}_{17}Cl$.

2   THE PERIODIC TABLE

The Periodic Table is an arrangement of the elements in ascending order of their atomic numbers so that elements with similar outer electronic configurations are in the same vertical column or group. There are 8 vertical groups and 7 horizontal periods. There are four pairs of elements for which their atomic numbers are the reverse of the order of their relative atomic masses. Thus, the properties of potassium put it in the same group as sodium and lithium although its relative atomic mass is less than that of argon. (For potassium $Z = 19$ and $Ar = 39·1$; for argon $Z = 18$ and $Ar = 39·95$.)

The Periodic Law first formulated by Mendeléev in 1869, **states that the properties of the elements are a periodic function of their relative atomic masses.**

A group in the Periodic Table is a vertical column of elements which have the same number of electrons in their outermost shells. Consider group I.

Element	Atomic number	Electronic configuration
H	1	1
Li	3	2,1
Na	11	2,8,1
K	19	2,8,8,1

1   **The group number of an element is numerically equal to the number of electrons in its outermost level** (except for the noble gases).

2   **The physical properties of the elements in a group change regularly down the group.** Consider F, Cl, Br and I in group I; atomic radii, melting points, boiling points and densities increase down the group.

3   **Elements in the same group have similar chemical properties because they have the same number of electrons in their outermost levels.** Thus, chlorine, bromine and iodine all have similar properties.

4   **The metallic properties of the elements in a group increase down a group;** as in for example the alkali metals in group I. In group IV, there is a transition from non-metals at the top (carbon and silicon) to metals at the bottom (tin and lead).

5   **The valency of an element may be the same as its group number.** This is seen in groups I to IV. Thus, sodium, magnesium, aluminium and silicon in groups I, II, III and IV respectively have valencies of 1, 2, 3 and 4 respectively. In groups IV to VII, the s- and p-electrons may behave as non-bonding (as inert or lone) pairs. Also s- and p-electrons may become unpaired so that the valency may rise in steps of two. The valencies in group III are 1 and 3, group IV (2 and 4), group V (1, 3 and 5), group IV (2, 4 and 6), group VII (1, 3, 5 and 7).

A period in the Periodic Table consists of a series of elements across the Periodic Table in which (excluding the transition elements), **the number of electrons increases from one to eight.**

1   **There is a gradual change from metallic to non-metallic character across a period.** There is an increase in electronegativity so that lithium is electropositive and fluorine is electronegative.

2   **There is a gradual change in the properties of the elements in a period.**

3   In a **short period** the number of electrons increases as the group number increases, the increase occurring in the outer s-shell only. The valency rises to four in the middle of the period and then decreases to one. In period 3, the maximum valency may be seven. Moving across a period from left to right, the metallic character of the elements decreases and their non-metallic character increases.

4   In a **long period**, the number of electrons increases successively first in the outer s-shell (two s-block elements), then in the penultimate d-shell (ten d-block elements) and lastly in the outer p-shell (six p-block elements). There is a decrease in the metallic properties from left to right across the period.

5   **s-block elements** form a block of reactive metals in groups I and II in which the outermost electrons are in the s-shell.

6   **p-block elements** are those elements in groups III, IV, V, VI, VII and O whose chemical properties are determined by the behaviour of the outer p electrons.

7   **d-block elements** (sometimes called transition elements) in periods 4, 5 and 6 are mostly metallic in character but they are less reactive than the s-block elements. The number of electrons in the outermost shell is constant across the period as the number of electrons in the penultimate d-shell increases (chromium and copper differ from this pattern). These elements show similarities across the period as well as their group characteristics. They generally show variable valencies.

8 f-block elements (or lanthanides and actinides) form a block of elements within the transition metals. They occur as a result of the successive addition of electrons to the f-subshells. There are 14 elements from cerium (Ce) to lutetium (Lu) beginning immediately after lanthanum. The properties across the f-block elements are so similar that it was difficult to separate the elements from each other.

9 Noble gases in group O have s- and p-shells filled.

Periodic properties

Properties which show periodic variation include:

1 Atomic volume (molar mass of the atoms of the element divided by density in $g\ cm^{-3}$) e.g. biggest values occur for the alkali metals of group I and the lowest values are shown by the elements of group IV or d-block elements.

2 Enthalpy of ionization and the noble gases have the largest values.

3 Metallic and non-metallic characteristics.

4 Valency.

Valency

Atoms react together chemically by making or breaking electron bonds.

1 Electrovalency occurs when one or more electrons are transferred from one atom to another forming ions e.g.

$$Na\ +\ Cl\ \longrightarrow\ Na^+\ Cl^-$$
$$2,8,1\quad 2,8,7\qquad\qquad 2,8\quad 2,8,8$$

2 Covalency occurs when two atoms share two or more electrons. The sharing may be equal as in the hydrogen molecule

$$H\cdot\ +\ H\cdot\ \longrightarrow\ H:H\quad or\quad H{-}H$$

or unequal as in a molecule of hydrogen chloride

$$H\cdot\ +\ \cdot\ddot{\underset{\cdot\cdot}{C}}l:\ \longrightarrow\ H^{\delta+}{-}Cl^{\delta-}$$

In **double bonds**, four electrons are shared and in **triple bonds**, six electrons are shared as in

$$H_2C{=}CH_2\quad and\quad HC{\equiv}CH$$
$$\text{ethene}\qquad\qquad\quad \text{ethyne}$$

3 Dative covalency (co-ordinate linkage) occurs when the two electrons in a covalent bond come from the same atom, as in the formation of an ammonium ion when the lone pair on the nitrogen is used.

$$H_3N: \; + H^+ \longrightarrow [H_3N \rightarrow H]^+ \quad \text{or} \quad [NH_4]^+$$

Metals and non-metals

Metals	Non-metals
1 Form positive ions by losing one or more electrons.	Form negative ions by gaining one or more electrons.
2 Replace hydrogen from acids forming salts.	Do not form salts in this way.
3 Form basic oxides which are alkalis if soluble in water.	Form acidic oxides which form acids in water.
4 Form few stable hydrides (Na, K and Ca hydrides contain H^-).	Form many stable hydrides.
5 Are good conductors of heat and electricity. Have high tensile strength, can be polished and are ductile.	Are bad conductors of heat and electricity (carbon conducts electricity), brittle, cannot be polished and are not malleable or ductile.

Metalloids are those elements which show properties characteristic of metals and non-metals, they may have high melting points and boiling points, high density but poor conductivity.

Metallic bond The outermost electrons in metal atoms move freely throughout the metal lattice. Thus, a metal consists of positive ions surrounded by a sea of moving electrons. The atoms are packed closely together in layers and because there are no rigid, directional bonds, the layers can slip.

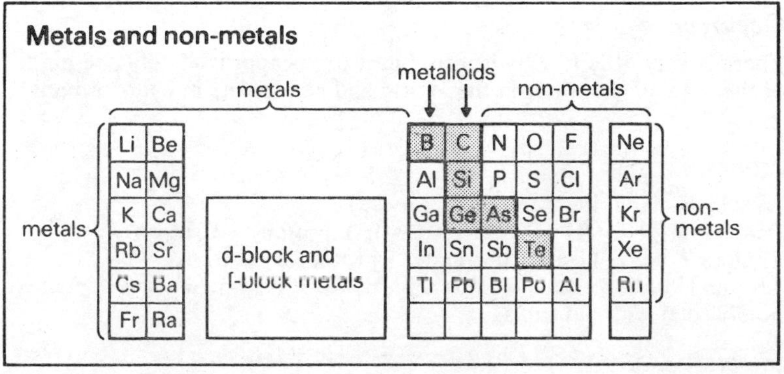

Practice questions

1   The physical and chemical properties of the elements are periodic functions of their atomic numbers. Explain why this is a more satisfactory basis for the modern Periodic Table than Mendeléev's use of atomic mass.

2   What is a group in the Periodic Table? Give five characteristics of the elements forming a group.

3   What is a period in the Periodic Table?

4   What are d-block elements? Outline their characteristics and explain how these properties are related to their electronic configurations. Illustrate your answer with reference to the elements scandium to zinc.

5   What are f-block elements? Illustrate your answer with reference to lanthanides.

6   What is meant by   (a) electrovalency,   (b) covalency?

7   Outline the properties of   (a) an electrovalent compound,   (b) a covalent compound.

8   What are the main physical and chemical properties by which metals differ from non-metals?
Explain how metallic properties of an element are related to its atomic structure and to its position in the Periodic Table.

9   Describe the distribution of electrons in the bonding in (a) sodium chloride,   (b) carbon dioxide,   (c) ammonium chloride.

10   Give the formulae of the chlorides of elements with atomic numbers 6 and 12. How do their electronic structures account for three of their differences in properties?

3   s-BLOCK ELEMENTS – HYDROGEN, H_2

Occurrence

There is very little free hydrogen, but it forms approximately one-ninth of the mass of the water in the World and also occurs in hydrocarbons.

Isotopes

Hydrogen ^1_1H (1 electron $(1s^1)$ 1 proton)
Deuterium D or ^2_1H (1 electron $(1s^1)$, 1 proton, 1 neutron)
Tritium T or ^3_1H (1 electrons $(1s^1)$, 1 proton, 2 neutrons)
Tritium is radioactive and emits β particles. It is formed when lithium is bombarded with neutrons.

$$^6_3Li + {}^1_0n \longrightarrow {}^4_2He + {}^3_1H.$$

Preparation

1 **Metal and non-oxidizing acid** such as zinc and dilute sulphuric acid in the cold. Reaction is slow with pure zinc and is catalysed by copper(II) sulphate solution. Hydrogen is collected over water or by upward displacement of air.

$$Zn(s) + H_2SO_4(aq) \longrightarrow ZnSO_4(aq) + H_2(g)$$

2 **Metal and alkali** such as zinc or aluminium with hot, concentrated sodium hydroxide solution.

$$Zn(s) + 2NaOH(aq) \longrightarrow ZnNa_2O_2(aq) + H_2(g)$$

3 **Metal and steam** Group I or II metals with water or steam e.g.

$$Mg(s) + H_2O(g) \longrightarrow MgO(s) + H_2(g)$$

4 **Electrolysis of water** acidified with a little dilute sulphuric acid – hydrogen is formed at the cathode and oxygen at the anode in the ratio 2:1 by volume.

5 **Metal hydride and water** such as lithium hydride and water.

$$LiH(s) + H_2O(l) \longrightarrow LiOH(aq) + H_2(g)$$

Manufacture

1 **From methane and petroleum** Methane or the naphtha fraction from the distillation of crude oil is heated with steam at 1200 K, at a pressure of 30 atmospheres, in the presence of nickel as a catalyst, e.g.

$$CH_4(g) + H_2O(g) \longrightarrow CO(g) + 3H_2(g)$$

The gaseous mixture is heated with excess steam at 700 K and 30 atmospheres in the presence of finely divided iron.

$$CO(g) + 3H_2(g) + H_2O(g) \longrightarrow CO_2(g) + 4H_2(g)$$

The carbon dioxide is dissolved in alkali or water under pressure and any traces of carbon monoxide are dissolved in ammoniacal copper(I) methanoate solution.

2 **Bosch process** Steam is passed over coke at 1300 K to form a mixture of carbon monoxide and hydrogen (water gas).

$$C(s) + H_2O(g) \rightleftharpoons CO(g) + H_2(g)$$

The water gas is mixed with more steam and passed over a catalyst of iron(III) oxide, Fe_2O_3, with chromium(III) oxide as a promoter.

$$CO(g) + H_2(g) + H_2O(g) \longrightarrow CO_2(g) + 2H_2(g)$$

The carbon dioxide and unchanged carbon monoxide are removed as above.

3   **Electroysis**   Hydrogen is obtained as a by-product of the electrolysis of brine in the manufacture of chlorine and sodium hydroxide.

Uses

1   Manufacture of ammonia – see Haber process in section on nitrogen.
2   Manufacture of margarine from liquid unsaturated esters.
3   Hydrogenation of coal to yield petrol and lubricating oil.
4   Manufacture of methanol. Water gas is mixed with more hydrogen and passed over a catalyst of zinc and chromium(III) oxides at 720 K and 200 atmospheres.

$$CO(g) + 2H_2(g) \longrightarrow CH_3OH(l)$$

5   Manufacture of hydrochloric acid – see section on halogens.
6   As a fuel, including rocket fuel.
7   Reduction of oxides to give free metals.
8   A future use may be the controlled fusion of hydrogen atoms at very high temperatures with the production of much energy.

Properties

Physical properties

Colourless, odourless, tasteless gas, very slightly soluble in water, soluble in some metals. It is the lightest gas and diffuses rapidly. It is a good conductor of heat and may be used as a coolant in a power station. Its low boiling and melting points (similar to noble gases) indicate the weakness of the forces between hydrogen molecules. Bond dissociation for the hydrogen-hydrogen bond is relatively high and the hydrogen molecule is very stable.

Ortho- and para-hydrogen   The two electrons in the hydrogen molecule are paired (Pauli's exclusion principle) but the two nuclei **may spin in the same direction** (ortho-hydrogen) **or in opposite directions** (para-hydrogen). At room temperature the ratio of o-hydrogen to p-hydrogen is 3:1 but near absolute zero, the amount of o-hydrogen is negligible.

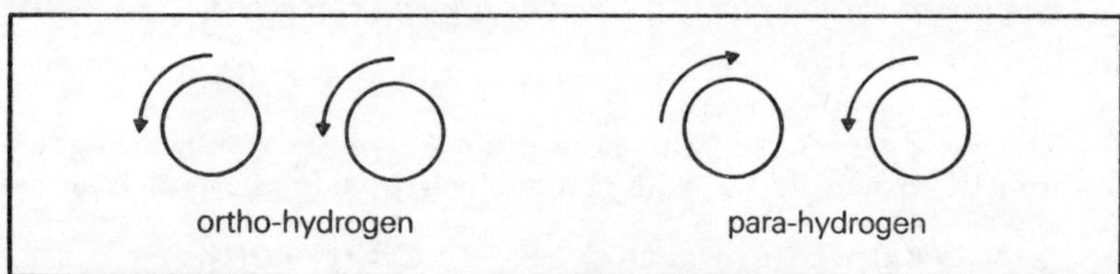

Chemical properties

Hydrogen forms more compounds than any other element. Its chemical behaviour is determined by its electronic configuration $1s^1$ and the small size of the atom.

1   It forms **electrovalent** compounds containing H^-.
2   It forms **electrovalent** compounds containing H^+.
3   It forms **covalent** compounds.
4   It forms **hydrogen bonds**.
5   It acts as a **reducing agent** and reduces the oxides of some d-block metals such as nickel, copper, lead and iron.
6   It '**adds on**' to unsaturated hydrocarbons, e.g. alkenes are reduced to alkanes in the presence of nickel.

Hydrogen compounds, hydrides

1   **Salt-like hydrides**   The s-block elements in group I and II form **ionic hydrides** in which the hydrogen atom accepts an electron, attaining a helium-like structure H^-. When, e.g. lithium, sodium, potassium and calcium are heated in a stream of dry hydrogen at 600 K, the hydride is formed with hydrogen acting as an **oxidizing agent**.

$$2Li(s) + H_2(g) \longrightarrow 2LiH(s)$$

These hydrides show a **crystal lattice** like that of **sodium chloride** but they are hydrolysed and react with water to form hydroxides. They are strong **reducing agents**.

$$2LiH(s) + H_2O(l) \longrightarrow 2LiOH(aq) + H_2(g)$$

These hydrides decompose when heated (LiH is more stable). When electrolysed, **hydrogen is formed at the anode**. H^- is a stronger **base** than OH^- and in liquid ammonia, H^- is a stronger base than NH_2^-.

$$H^- + NH_3(l) \longrightarrow NH_2^- + H_2(g)$$

2   **The positive hydrogen ion or proton, H^+**, is formed in **aqueous solutions** and is hydrated, e.g. H_3O^+. The production of the proton is characteristic of **acids** e.g.

$$HCl + aq \longrightarrow H^+(aq) + Cl^-(aq)$$

3   **Covalent hydrides**   These are numerous and hydrides of p-block elements are mainly **covalent**, e.g. B_2H_6; $(AlH_3)_n$. **Stability of the hydrides decreases** down a group as the size of the atom and its electropositive character increases, e.g. ammonia is less readily decomposed by heat than phosphine. Hydrogen combines directly with the **halogens**, the reaction with fluorine is explosive even in the dark, the reaction with chlorine is explosive in sunlight and the reaction with iodine is reversible (hydrogen iodide is formed at 400 K in the presence of platinised asbestos). Hydrogen reacts reversibly

with **nitrogen** (Haber process) and when pure, burns quietly in **oxygen** to form water (a mixture of hydrogen and oxygen is dangerously inflammable). Compounds are mostly gaseous at room temperature and pressure with weak intermolecular forces – see also hydrogen bonding.

4   **Hydrogen bonding This is a dipole-dipole attraction occurring between a hydrogen atom joined to a strongly electronegative atom and another electronegative atom with a lone pair of electrons.** The strongest are found in hydrogen fluoride. The hydrogen bond is stable in ice, accounting for its open structure so that for water, the density of the solid form is less than that of the liquid. Each water molecule is tetrahedrally surrounded by four other water molecules, the final structure depending on the temperature and pressure.

water

alcohols

monocarboxylic acids

ice

The hydrogen bond causes the boiling points of H_2O, HF and NH_3 to be higher than expected. [For further details see Rapid Revision Physical Chemistry.]

5   **Interstitial hydrides**   d-block elements may take up large volumes of hydrogen, the products having no simple formulae. Finely divided palladium may absorb up to 900 times its volume of gas which is expelled on heating. The hydrogen atoms seem to fit into spaces between the atoms of the other element and the density of the interstitial hydride is less than that of the corresponding metal.

6   **Complex hydrides**   Formed by transition metals and group III elements boron, aluminium and gallium e.g. [Li AlH_4], Na[BH_4],

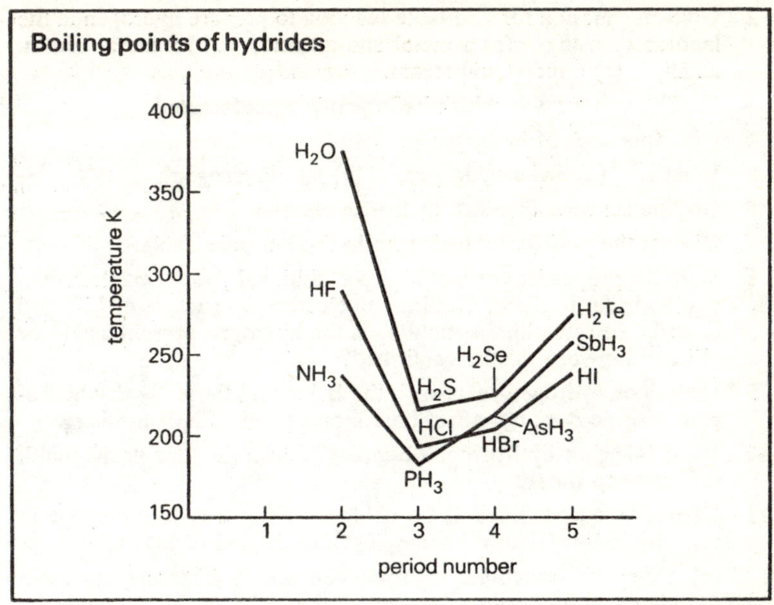

$Al[BH_4]_3$ and $Li[BH_4]$. The MH_4^- ion has a tetrahedral structure: they are reducing agents.

7 **Atomic hydrogen** Hydrogen is 60% dissociated when the gas is blown through an electric arc between tungsten electrodes. A jet of atomic hydrogen liberates a lot of energy as the atoms recombine.

Anomalous position of hydrogen in the Periodic Table

1 Properties determined by the small size of the atom.
2 Electronegativity does not correspond to the value expected for elements in group I or II.
3 Hydrogen shows similarities with the alkali metals.
4 Hydrogen shows similarities with the halogens.
5 Hydrogen cannot be identified rigidly in either group and occupies a unique position at the top of the Periodic Table.

Practice questions

1 Name the three isotopes of hydrogen. How do they differ in
   (a) electronic,   (b) nuclear structure?

2   Give an equation for a suitable reaction to prepare hydrogen in the laboratory using   (a) a metal and an acid,   (b) a metal and an alkali,   (c) a metal and steam.

3   Outline two ways in which hydrogen is manufactured.

4   Give four uses of hydrogen.

5   What is   (a) *ortho*-hydrogen,   (b) *para*-hydrogen?

6   Outline six ways in which hydrogen can react.

7   Discuss the position of hydrogen in the Periodic Table.

8   State the properties you would expect lithium hydride, an electrovalent hydride, to show. Outline the electronic structure of LiH and describe its crystalline structure. Is the hydrogen present as H^+ or H^- and how can this be confirmed?

9   How does hydrogen react with the halogens? Perfectly dry hydrogen chloride does not affect blue litmus paper. Explain this.

10   What is meant by hydrogen bonding? Illustrate your answer with reference to water.

11   Hydrogen may behave as an oxidizing or as a reducing agent to form hydrides. Give two examples of each kind of reaction.

12   (a) Give five reactions of hydrogen which illustrate its main chemical properties. Give two industrial uses.

    (b) Explain the bonding in each of the following: sodium hydride, hydrogen chloride, methane, potassium hydrogen fluoride, lithium aluminium hydride.

4   p-BLOCK ELEMENTS: NOBLE (RARE OR INERT) GASES OF GROUP O

The noble gases (helium, neon, argon, krypton, xenon and radon) have an outer electronic configuration of ns^2np^6 (helium $1s^2$).

He	$1s^2$						
Ne	$1s^2$ $2s^2$	$2p^6$					
Ar	$1s^2$ $2s^2$	$2p^6$ $3s^2$ $3p^6$					
Kr	$1s^2$ $2s^2$	$2p^6$ $3s^2$ $3p^6$ $3d^{10}$	$4s^2$ $4p^6$				
Xe	krypton core	$4d^{10}$	$5s^2$ $5p^6$				
Rn		$4d^{10}$ $4f^{14}$ $5s^2$ $5s^2$ $5p^6$ $5d^{10}$	$6s^2$ $6p^6$				

General properties

1   Colourless, odourless, tasteless gases, found in the atmosphere.

2   Obtained by the fractional distillation of liquid air.

3   Once thought that they only formed clathrate compounds in which the gaseous atoms were trapped within the crystal lattices of other compounds.
4   Helium diffuses through quartz.
5   Platinum fluoride oxidizes xenon forming $Xe^+[PtF_6]^-$.
6   XeF_4 and XeF_2 also formed.
7   In XeF_2, two electrons from the fluorine atoms are shared with the eight electrons of the xenon atom. The molecule is linear.
8   In XeF_4, four electrons from the fluorine atoms are shared with the xenon atom. Molecule is square planar in shape.
9   Number of oxides formed – XeO_3, XeO_4^{2-}, XeO_4, XeO_6^{4-}.
10   Helium filled the first balloon to fly the Atlantic; mixed with oxygen to form a breathing mixture for deep-sea divers; liquid helium cools materials that are superconductors.
11   Neon – gives a bright orange-red glow in discharge lamps, used as a starter gas in sodium lamps.
12   Argon – fills electric light bulbs; provides an inert atmosphere in the extraction of elements like titanium, used in miners' head-lamps and in the lights of lighthouses.

Practice questions

1   Suggest reasons why the noble gases show very little chemical activity.

2   Give four uses of the noble gases and state why they are so used.

3   What is a clathrate compound? Why are the noble gases able to form such compounds?

5   s-BLOCK ELEMENTS: THE ALKALI METALS OF GROUP I

Lithium, sodium, potassium, rubidium and caesium all have an outer electronic configuration of ns^1 and an oxidation state of $+1$.

Li $1s^2\ 2s^1$
Na $1s^2\ 2s^2\ 2p^6\ 3s^1$
K $1s^2\ 2s^2\ 2p^6\ 3s^2\ 3p^6\qquad 4s^1$
Rb $1s^2\ 2s^2\ 2p^6\ 3s^2\ 3p^6\ 3d^{10}\ 4s^2\ 4p^6\qquad 5s^1$

General properties

1   The relatively large size of the atoms and the single electron in the outer shell gives these elements low first ionization energies.

2   Reactivity and electropositive character increase down the group as the relative atomic masses increase: the outer electronic shell is shielded from the nucleus by inner complete shells and the outer electron is lost with increasing ease as atomic size increases.

3   Because the outer electron is easily excited, these elements impart characteristic colours to the Bunsen burner flame, e.g. sodium – bright yellow, potassium – lilac.

4   Because of their high reactivity, the metals are stored under oil, away from air.

5   Melting points decrease down the group, weak inter-atomic forces.

6   Metals are soft and can be cut with a knife (not lithium).

7   Elements react with water to form hydrogen and hydroxides.

8   All form basic oxides $(M^+)_2O^{2-}$ and hydroxides, M^+OH^-, which dissolve in water to form alkalis. All form peroxides $(M^+)_2(O{-}O)^{2-}$ and superoxides $M^+(O{-\cdot\cdot-}O)^-$.

9   All form hydrides, M^+H^-, which form hydrogen with water. Stability of hydrides decreases down the group.

10 Most of their compounds, e.g. hydroxides, carbonates and sulphates are stable to heat; nitrates form nitrites; ease of thermal decomposition of nitrates decreases down the group.

11 Salts are white unless the anion is coloured.

12 Salts of the metals are nearly all soluble in water.

13 Because of their large size and small charge, these elements have little tendency to form complexes.

14 These elements are amongst the strongest reducing agents.

15 All are good conductors of heat and electricity.

Lithium, Li

Because of its small size and greater electronegativity, lithium differs in several ways from the other alkali metals. In some ways it resembles the elements of group II and has a diagonal relationship with magnesium.

1   Lithium chloride, LiCl, is very soluble in water and is slightly hydrolysed. It is more soluble than the chlorides of potassium and sodium. It is also deliquescent like calcium chloride and forms a hydrate $LiCl\cdot H_2O$.

2   Lithium carbonate is sparingly soluble in water like magnesium and calcium carbonates, the carbonates of sodium and potassium are very soluble. Lithium carbonate is not decomposed by heat.

3   Lithium does not form a hydrogencarbonate.

4   Lithium phosphate is almost insoluble.

5   Lithium fluoride is almost insoluble, like calcium fluoride.

6   Virtually all lithium salts are hydrated. Hydration decreases down the group.

7   Lithium forms only the monoxide, Li_2O.

8 Lithium nitrate decomposes when heated to form the oxide. It crystallizes as $LiNO_3.3H_2O$.

Sodium, Na

Occurrence Sodium chloride in salt mines and sea water; Chile – saltpetre (**sodium nitrate**), soda (**sodium carbonate**), **borax** ($Na_2B_4O_7.10H_2O$) and **cryolite** (Na_3AlF_6).

Extraction Sodium is obtained by the **electrolysis of fused sodium chloride** obtained from salt mines or by the evaporation of sea water. Calcium chloride is added to lower the melting point of the electrolyte (from 1074 K to 870 K). The **Down's cell** has a steel casing lined with refractory brick, a steel cathode and a carbon (graphite) anode surrounded by an iron diaphragm. The electrode potential is greater than would be expected because a film of chlorine gas builds up on the anode, causing polarization. Resistance to the current produces sufficient heat to keep the electrolyte molten.

At anode (+) $\qquad 2Cl^- \longrightarrow Cl_2(g) + 2e^-$

At cathode (−) $\qquad Na^+ + e^- \longrightarrow Na$

Uses
1 **In alloys**, e.g. tetraetnyl-lead(IV) used as an anti-knock in petrol.
2 **As a source of sodium compounds**, e.g. sodium peroxide.
3 **As a reducing agent**, e.g. to reduce titanium(IV) chloride to titanium; in Wurtz and Fittig reaction and in Lassaigne's test for carbon, sulphur and halogens in organic chemistry. Sodium-sulphur may be used in a fuel cell.
4 **As a conductor** e.g. for the transference of heat at Dounreay.

Physical properties
Sodium has only one isotope. Very soft, silvery metal, good conductor of heat and electricity, dissolves in mercury forming an amalgam.

Chemical properties
1 **Very reactive, oxidation state of +1**, stored under oil.
2 **In moist air**, it forms in succession Na_2O, $NaOH$, $NaOH(aq)$ and finally $Na_2CO_3.10H_2O$ which may effloresce to $Na_2CO_3.H_2O$.
3 **On heating in air** forms monoxide and peroxide.

$$4Na(s) + O_2(g) \longrightarrow 2Na_2O(s)$$

$$2Na(s) + O_2(g) \longrightarrow Na_2O_2(s)$$

4 **With oxygen**, forms Na_2O, and with excess oxygen forms Na_2O_2. **Sodium peroxide** is obtained by passing carbon dioxide-free air over trays of sodium.

Practice questions

1   The general electronic configurations for the elements in group I of the Periodic Table is ns^1. Write out the electronic configurations for $_3Li$, $_{11}Na$, $_{19}K$, $_{37}Rb$.

2   With reference to the alkali metals of group I, comment on (a) their first ionization energies and reactivity,   (b) their effect on a Bunsen burner flame,   (c) their tendency to form coloured ions and complexes.

3   The crystal structure of the halides of the alkali metals changes down the group. Suggest a reason for this.

4   How and why does lithium differ from the other alkali metals in group I?

5   How is sodium obtained commercially? Give three uses of the metal.

6   How does sodium react with   (a) oxygen,   (b) hydrogen, (c) water,   (d) ammonia?

7   How is sodium hydroxide produced commercially? How does it react with   (a) iron(II) sulphate,   (b) zinc metal,   (c) ammonium chloride,   (d) dilute sulphuric acid? Give equations for all reactions.

8   How does sodium hydroxide solution react with   (a) silicon, (b) white phosphorus,   (c) red phosphorus,   (d) chlorine? Give equations for all reactions.

9   How is sodium carbonate manufactured by the ammonia-soda process? Explain the importance of temperature control throughout. In what ways is this process efficient? Why is a similar process impracticable for making potassium carbonate?

10   How would you obtain precipitates of   (a) calcium carbonate, (b) zinc carbonate,   (c) copper carbonate?

11   Give equations for the effect of heat (if any) on   (a) potassium carbonate,   (b) lithium carbonate,   (c) sodium hydrogen-carbonate,   (d) sodium nitrate,   (e) sodium sulphate.

12   Explain why group I elements are   (a) univalent,   (b) largely ionic in combination,   (c) strong reducing agents,   (d) poor complexing agents.

13   Describe the principal features of the periodic classification of the elements. Explain what is meant by a group, illustrating your answer with reference to the alkali metals of group I.

14   Compare and contrast the electronic structures and properties of potassium and sodium. Show why they are in the same group of the Periodic Table.

5 **With halogens**, on warming, combines directly forming **halides**, Na^+Y^-.

6 **With sulphur** and **hydrogen**, combines directly forming Na_2S and NaH respectively. Forms a polysulphide with excess sulphur.

7 **With water**, forms sodium hydroxide and hydrogen.

$$2Na(s) + 2H_2O(l) \longrightarrow 2NaOH(aq) + H_2(g)$$

8 **With ammonia**, dissolves to form a blue solution. **Sodamide** $NaNH_2$, is formed when sodium is heated in dry ammonia. Sodamide is hydrolysed by water.

$$2Na(s) + 2NH_3(g) \longrightarrow 2NaNH_2(l) + H_2(g)$$

$$NaNH_2(s) + H_2O(l) \longrightarrow NaOH(aq) + NH_3(g)$$

9 **Reduces the chlorides or oxides** of less electropositive metals on heating e.g. aluminium chloride.

$$AlCl_3(s) + 3Na(s) \longrightarrow 3NaCl(s) + Al(s)$$

10 **Reduces carbon dioxide** on heating,

$$4Na(s) + 3CO_2(g) \longrightarrow 2Na_2CO_3(s) + C(s)$$

Compounds of sodium

Sodium oxide $(Na^+)_2O^{2-}$, is best obtained by heating sodium in a limited amount of oxygen and distilling off excess sodium *in vacuo*. Basic oxide reacts violently with acids to give salts and water, also forms sodium hydroxide with water.

Sodium peroxide, $(Na^+)_2(O—O)^{2-}$, is white when absolutely pure but is usually pale yellow. Forms hydrogen peroxide with cold dilute acids and reacts with carbon dioxide to liberate oxygen – used for purifying air.

$$Na_2O_2(s) + H_2SO_4(aq) \longrightarrow Na_2SO_4(aq) + H_2O_2(aq)$$

$$2Na_2O_2(s) + 2CO_2(g) \longrightarrow 2Na_2CO_3(s) + O_2(g)$$

Very strong oxidizing agent, e.g. changes green chromium(III) salts to yellow chromium(VI) and sulphides to sulphates.

Sodium superoxide $Na^+(O—\!\cdot\!-\!\cdot\!-O)^-$ contains a three-electron bond. [This bond is formed only between atoms of the same or similar electronegativities and is about half as strong as a covalent bond. It may be considered as a resonance hybrid between $A\cdot:B$ and $A:\cdot B$]. Paramagnetic and coloured yellow (KO_2, yellow; RbO_2, orange; CsO_2, red). Very strong oxidizing agent.

Sodium hydroxide, Na^+OH^-

Brine passes over a **flowing mercury cathode**. The carbon anode dips into the brine, the distance between the electrodes being about 2 mm.

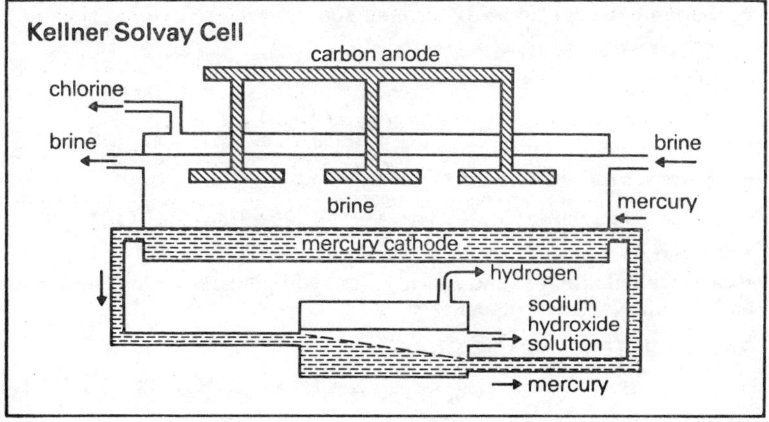

At anode (+) $2Cl^- \longrightarrow 2e^- + Cl_2(g)$

At cathode (−) $Hg + Na^+ + e^- \longrightarrow Na/Hg$

Hydrogen does not discharge at the mercury surface because of a high overvoltage. The **sodium amalgam** flows into a lower cell containing **distilled water** in contact with steel. Hydrogen is discharged and sodium hydroxide formed. The solution is evaporated giving solid sodium hydroxide. This is also the main source of chlorine.

$$2Na/Hg + 2H_2O(l) \longrightarrow 2(Na^+OH^-)aq + H_2(g) + 2Hg(l)$$

In the **Gibbs cell**, the carbon anode and iron cathode are separated by a porous asbestos diaphragm. Hydrogen discharges at the cathode and water ionizes leaving an excess of OH^- to form sodium hydroxide. Chlorine is discharged at the anode.

Properties

As an alkali
1   White, slippery to the touch, solid, causes caustic burns.
2   Very soluble in water with the evolution of much heat.
3   Fully ionized and precipitates insoluble hydroxides of metals,

$$Fe^{3+}(aq) + 3OH^-(aq) \longrightarrow Fe(OH)_3(s)$$

4   Hydroxides may be amphoteric and re-dissolve,

$$Zn(OH)_2(s) + 2OH^-(aq) \longrightarrow [Zn(OH)_4]^{2-}(aq)$$

5   Liberates ammonia when heated with ammonium salts,

$$NH_4^+(aq) + OH^-(aq) \longrightarrow NH_3(g) + H_2O(l)$$

6　Saponifies organic acids forming soaps,

$$CH_3.COOC_2H_5(l) + NaOH(aq) \longrightarrow$$
ethyl ethanoate

$$CH_3COO^-Na^+(aq) + C_2H_5OH(l)$$
sodium ethanoate　　　　ethanol

7　Reacts with acids to form salts and water,

$$NaOH(aq) + HCl(aq) \longrightarrow NaCl(aq) + H_2O(l)$$

With metals

Reacts with **aluminium** and **zinc** to form hydrogen and sodium aluminate and sodium zincate respectively.

$$2Al(s) + 2OH^-(aq) + 10H_2O(l) \longrightarrow$$

$$2[Al(H_2O)_2(OH)_4]^{2-}(aq) + 3H_2(g)$$

$$Zn(s) + 2OH^-(aq) + 2H_2O(l) \longrightarrow [Zn(OH)_4]^{2-}(aq) + H_2(g)$$

With non-metals

1　With **silicon** forms hydrogen and sodium silicate(IV) on heating.

$$Si(s) + 2OH^-(aq) + H_2O(l) \longrightarrow SiO_3^{2-}(aq) + 2H_2(g)$$

2　With **white phosphorus** forms phosphine and sodium phosphinate.

$$3OH^-(aq) + P_4(s) + 3H_2O(l) \longrightarrow 3H_2PO_2^-(aq) + PH_3(g)$$

Red phosphorus does not react with sodium hydroxide.

3　With **cold sodium hydroxide solution, chlorine** forms sodium chloride and sodium chlorate(I).

$$2OH^-(aq) + Cl_2(g) \longrightarrow Cl^-(aq) + OCl^-(aq) + H_2O(l)$$

With **hot concentrated** sodium hydroxide solution, chlorine forms sodium chloride and sodium chlorate(V).

$$6OH^-(aq) + 3Cl_2(g) \longrightarrow 5Cl^-(aq) + ClO_3^-(aq) + 3H_2O(l)$$

Uses

Used in the extraction of phenols and methylphenols from tar, saponification in the manufacture of paper, artificial silk and dye stuffs.

Sodium carbonate, soda ash Na_2CO_3, washing soda $Na_2CO_3.10H_2O$
Manufacture – ammonia-soda or Solvay process.
Ammonia gas is passsed into an almost saturated solution of **sodium chloride**. The temperature rises but the liquid is cooled to 310 K and then passed down a carbonating or Solvay tower up which **carbon dioxide** rises. The reaction is exothermic and the tower must be cooled (from approximately 340 K to 300 K) and several towers are used.

$$2NH_3(g) + CO_2(g) + H_2O(l) \longrightarrow 2NH_4^+(aq) + CO_2^{2-}(aq)$$

$$CO_3^{2-}(aq) + CO_2(g) + H_2O(l) \longrightarrow 2HCO_3^-(aq)$$

The least soluble ions precipitate out as **sodium hydrogencarbonate** which is removed by vacuum filtration and washed free of ammonium compounds. It is then heated to form **sodium carbonate** and recrystallized from water to form $Na_2CO_3.10H_2O$.

$$2NaHCO_3(s) \longrightarrow Na_2CO_3(s) + CO_2(g) + H_2O(l)$$

The carbon dioxide is returned to the tower.

1 **Brine** is obtained from natural common salt deposits.
2 **Carbon dioxide** is obtained by heating limestone

$$CaCO_3(s) \rightleftharpoons CaO(s) + CO_2(g)$$

3 **Ammonia** is regenerated by steam heating the ammonium chloride produced in the process with lime.

$$CaO(s) + H_2O(l) \longrightarrow Ca(OH)_2$$

$$2NH_4Cl(aq) + Ca(OH)_2(s) \longrightarrow$$
$$CaCl_2(aq) + 2H_2O(l) + 2NH_3(g)$$

In practice a little ammonia has to be added to replace losses. This process is possible because of the low solubility of sodium hydrogencarbonate. A similar process is not possible for potassium compounds because potassium hydrogencarbonate is too soluble.

Uses
1 Manufacture of glass.
2 Water softener (as washing soda).

$$Ca^{2+}(aq) + CO_3^{2-}(aq) \longrightarrow CaCO_3(s)$$

3 In paper making.
4 In detergents and cleansers.

Properties
1 Crystallizes as $Na_2CO_3.10H_2O$, between $305.2\,K$ and $308.6\,K$, $Na_2CO_3.7H_2O$ can be crystallized. Shows efflorescence forming $Na_2CO_3.H_2O$.
2 Hydrolysed in solution, $pH > 7$ and solution is alkaline.

$$CO_3^{2-} + H_2O(l) \rightleftharpoons HCO_3^-(aq) + OH^-(aq)$$

3 Forms carbon dioxide and sodium salts with mineral acids.

$$2H^+(aq) + CO_3^{2-}(aq) \longrightarrow H_2O(l) + CO_2(g)$$

4 Precipitates carbonates (except those of potassium and ammonium), e.g.

$$Ca^{2+}(aq) + CO_3^{2-}(aq) \longrightarrow CaCO_3(s)$$

Basic carbonate may be precipitated, e.g. for magnesium, copper, zinc and lead – use sodium hydrogencarbonate.

$$2Mg^{2+}(aq) + 2OH^-(aq) + CO_3^{2-}(aq) \longrightarrow MgCO_3.Mg(OH)_2(s)$$

$$Mg^{2+}(aq) + 2HCO_3^-(aq) \longrightarrow MgCO_3(s) + H_2O(l) + CO_2(g)$$

Sodium hydrogencarbonate, $NaHCO_3$
Manufacture, see sodium carbonate.

Uses
1 In baking powder when mixed with sodium dihydrogenphosphate and rice powder. The carbon dioxide formed makes cakes 'light'.
2 In health salts mixed with a weak acid, e.g. sodium tartrate or acid sodium phosphate(V).
3 In fire extinguishers.

Properties
1 With the exception of lithium, the alkali metals are the only elements able to form stable hydrogen carbonates.
2 Decomposes on heating.

$$2NaHCO_3(s) \longrightarrow Na_2CO_3(s) + H_2O(l) + CO_2(g)$$

3 Liberates carbon dioxide with dilute acids.
4 Hydrolysed by water to yield an alkaline solution – less hydrolysed than normal carbonate.

$$HCO_3^- + H_2O(l) \rightleftharpoons H_2CO_3(aq) + OH^-(aq)$$

5 Hydrogen bonding is important in the structure of the crystal lattice.

Sodium chloride, common salt, $NaCl$
Manufacture Rock salt is mined and salt is also obtained by the evaporation of sea water.

Uses
1 In the flavouring of most foods and also in preserving, e.g. curing bacon and pickling onions.
2 Manufacture of soap.
3 As a glaze for earthenware, e.g. drain-pipes.
4 Electrolysis of brine gives sodium hydroxide and chlorine.
5 Electrolysis of fused sodium chloride gives sodium.
6 Used in ammonia-soda process to form sodium carbonate.

Properties White, ionic solid, fairly soluble in water, but solubility changes little with temperature; solid has a face-centred cubic structure (crystal structure changes from face-centred cubic as the size of the metallic ion increases, RbCl, face-centred primitive and CsCl, primitive cubic).

Other sodium compounds
Sodium nitrate, $NaNO_3$, occurs as Chile saltpetre. Very soluble in water, commercial product is deliquescent, on heating forms sodium nitrite.

$$2NaNO_3(s) \longrightarrow 2NaNO_2(s) + O_2(g)$$

Used in the manufacture of nitric acid and potassium nitrate and also as a nitrogenous fertilizer.

Sodium nitrite, $NaNO_2$, is used in the manufacture of nitrous acid and in the diazotization of aromatic amines.

Sodium sulphate, Na_2SO_4, is made commercially by the action of common salt on concentrated sulphuric acid at red-heat. Used in the manufacture of glass, water-glass, sodium sulphide, sodium thiosulphate and as Glauber's salt. The solubility curve of sodium sulphate shows a break at 305·5 K. The solubility reaches a maximum and then decreases with the anhydrous salt separating out as a fine powder.

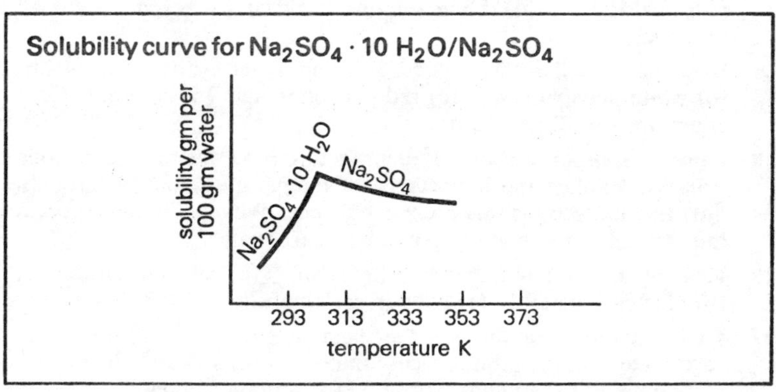

Sodamide, $NaNH_2$, is obtained when sodium is heated in dry ammonia at 700 K.

$$2Na(l) + 2NH_3(g) \longrightarrow 2NaNH_2(l) + H_2(g)$$

Sodium cyanide, NaCN, is obtained when molten sodamide is run on to white hot charcoal,

$$2NaNH_2(l) + C(s) \longrightarrow Na_2CN_2(l) + 2H_2(g)$$
$$Na_2CN_2(l) + C(s) \longrightarrow 2NaCN(l)$$

It is used in the extraction of silver.

6 s-BLOCK ELEMENTS: ALKALINE EARTH METALS OF GROUP II

Beryllium, magnesium, calcium, strontium and barium all have an outer electronic configuration of ns^2 and an oxidation state of $+2$.

Be $1s^2\ 2s^2$
Mg $1s^2\ 2s^2\ 2p^6\ 3s^2$
Ca $1s^2\ 2s^2\ 2p^6\ 3s^2\ 3p^6 \qquad 4s^2$
Sr $\boxed{1s^2\ 2s^2\ 2p^6\ 3s^2\ 3p^6\ 3d^{10}\ 4s^2\ 4p^6} \qquad 5s^2$
Ba $\boxed{1s^2\ 2s^2\ 2p^6\ 3s^2\ 3p^6\ 3d^{10}\ 4s^2\ 4p^6}\ 4d^{10}\ 5s^2\ 5p^6\ 6s^2$

krypton core

General properties

1 Properties dominated by the two outermost electrons of each element.
2 Because of the increasing nuclear charge, the atoms of the alkaline earth metals are smaller than those of the alkali metals.
3 Relatively poor screening effect of the d-electrons enables the outer electrons to be pulled in and held more firmly causing strontium, barium and radium to have higher densities than those of beryllium, magnesium and calcium which do not have any d-electrons.
4 The chemistry of calcium, strontium and barium is essentially that of the corresponding M^{2+} ions and similar to that of the alkali metals.
5 The first ionization energies of the group II elements are greater than those of group I and the second ionization energies are even greater. This is offset by other energy changes which favour the formation of the divalent ion: in crystals favourable lattice energies and in solution, favourable hydration energies.
6 Because the ions are smaller and the charge is higher than on the alkali metals, the alkaline earth metals show a greater degree of hydration e.g. $MgCl_2.6H_2O$, $CaCl_2.6H_2O$ and $BaCl_2.2H_2O$.
7 M^{2+} ions are all diamagnetic and compounds are colourless unless the anions are coloured.
8 The ions colour the Bunsen burner flame – calcium (brick-red), strontium (crimson), barium (apple-green).
9 The two electrons available for metallic bonding (compared with one per atom for the alkali metals) make the alkaline earth metals harder than the alkali metals and also give the alkaline earth metals higher melting and boiling points.
10 The irregular melting and boiling points can be explained by different metallic crystal structures.

11   Down the group, ionic size increases so that hydroxides are stronger bases, salts are less hydrated and usually more soluble, carbonates are more stable.

12   Down the group, ionization energy decreases so that electropositive character increases and reactivity of the metal increases.

13   All metals are strongly electropositive and react with water (less readily than alkali metals, magnesium readily with steam and beryllium not at all).

14   All react with dilute acids forming hydrogen.

15   Salts of the metals do not hydrolyse, except for some of magnesium.

16   Form salt-like hydrides $M^{2+}(H^-)_2$ which form hydrogen with water – reducing agents.

17   All carbonates are insoluble in water and decomposed by heat.

18   Strontium and barium show no tendency to form complex ions, calcium and magnesium show a slight tendency.

19   Hydrogencarbonates are unstable in the solid state.

20   All the elements burn in oxygen to form white solid monoxides $M^{2+}O^{2-}$ which form hydroxides exothermically with water.

21   Thermal stability and solubility of the hydroxides increase with increasing electropositive character, from magnesium to barium, sparingly soluble, reasonably strong bases.

22   Halides formed by direct combination of metal and halogen or by treating carbonate with halogen acid. Hygroscopic and form hydrates which on further exposure to the atmosphere deliquesce. Anhydrous $CaCl_2$ used as a drying agent. Halides (except fluorides) are soluble.

23   Nitrates decompose on heating to form oxides.

24   Sulphates show varying degrees of hydration. Beryllium sulphate exists as $BeSO_4.4H_2O$ below 362 K, above 362 K $BeSO_4.2H_2O$ is stable. Magnesium sulphate has hydrates corresponding to 1, $1\frac{1}{4}$, $1\frac{1}{2}$, 4, 5, 6 and 7 molecules of water of crystallization. $MgSO_4.7H_2O$ exists in two allotropic forms – rhombic (Epsom salts) and monoclinic. Gypsum, $CaSO_4.2H_2O$ forms a hemi-hydrate or plaster of Paris $(CaSO_4)_2.H_2O$ at 398 K and the anhydrite at 413 to 473 K. Strontium and barium sulphates are anhydrous only. Solubility decreases down the group.

Beryllium, Be

Because of its small size, and relatively high electronegativity and ionization energy values, beryllium has **more covalent character** than the other alkaline earth metals. Because there are no d-orbitals, the **co-ordination number** of beryllium is **limited** and beryllium compounds never have more than **four molecules of water of crystallization**. It forms more complexes than other members of group II, e.g. $[BeF_4]^{2-}$.

1 Does not combine directly with hydrogen and forms a covalent hydride when beryllium chloride is reduced by an ethereal solution of lithium tetrahydride aluminate(III).

$$2BeCl_2 + LiAlH_4 \longrightarrow 2BeH_2 + LiCl + AlCl_3$$

Possible structure:

$$\left[\begin{array}{c} H \diagdown \quad \diagup H \diagdown \\ \quad Be \quad\quad Be \\ H \diagup \quad \diagdown H \diagup \end{array} \right]_n$$

2 Does not react with water or to any extent with steam, any product being the oxide not the hydroxide.

3 Reacts slowly with dilute acids.

4 Amphoteric, reacting with sodium and potassium hydroxides to form beryllates and hydrogen (*c.f.* aluminium).

5 Oxide and hydroxide are amphoteric.

6 Halides BeX_2 are covalent and fume in moist air due to hydrolysis.

7 Beryllium carbonate is unstable.

8 Combines directly with carbon in an electric furnace, forming Be_2C. The compound BeC_2 is obtained by reacting beryllium with ethyne.

Magnesium, Mg

Occurrence Magnesite, $MgCO_3$ and **dolomite**, $MgCO_3.CaCO_3$ and also as **sulphate, chloride** and **silicates**.

Extraction **Dow's process. Lime** is used to precipitate **magnesium hydroxide** from sea water. Precipitate is treated with **hydrochloric acid** and **anhydrous magnesium chloride** is obtained by evaporation in a stream of hydrogen chloride to prevent hydrolysis. The magnesium chloride is mixed with a little **sodium chloride** so that the temperature of the mixture when electrolysed is kept at 1000 K. Magnesium is discharged at the iron cathode and floats on the surface of the electrolyte. It is protected from chlorine by a porcelain sheath.

Magnesium may also be obtained by **chemical reduction. Dolomite** is heated to form a mixture of calcium and magnesium oxides which is then mixed with **ferrosilicon**. When this is heated electrically in a retort of chromium steel at 1500 K and at very low pressure, magnesium vaporizes and distils.

$$2MgO(s) + CaO(s) + Si(s) \longrightarrow CaSiO_3(l) + 2Mg(g)$$

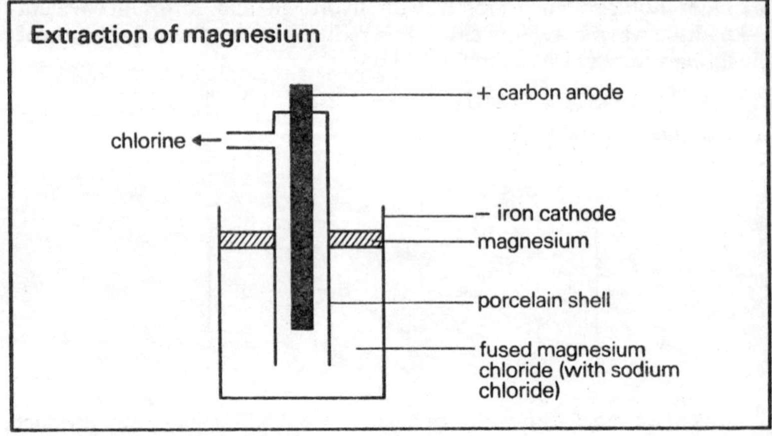

Uses
1   Mainly used in production of light alloys, e.g. duralumin, magnalium
     and elecktron.
2   Burns brilliantly, e.g. in fireworks and flares.
3   Grignard reagents used in synthesis of organic compounds.
4   Used in extraction of titanium.

Physical properties
Silvery, white metal.

Chemical properties
1   Does not react with dry air, oxidizes in damp air forming oxide which
     becomes hydroxide and then carbonate.
2   Burns to form oxide and nitride.

$$2Mg(s) + O_2(g) \longrightarrow 2MgO(s)$$

$$3Mg(s) + N_2(g) \longrightarrow Mg_3N_2(s)$$

3   Pure magnesium reacts very slowly with cold water but it burns
     brilliantly when heated in steam – reducing action.

$$Mg(s) + H_2O(g) \longrightarrow MgO(s) + H_2(g)$$

4   Dilute non-oxidizing acids form salts and hydrogen. Dilute nitric acid
     yields oxides of nitrogen.

$$3Mg(s) + 8HNO_3(aq) \longrightarrow 3Mg(NO_3)_2(g) + 4H_2O(l) + 2NO(g)$$

Hot concentrated sulphuric acid yields sulphur dioxide.

$$Mg(s) + 2H_2SO_4(l) \longrightarrow MgSO_4(aq) + 2H_2O(l) + SO_2(g)$$

5 Does not react with alkalis.

6 Combines with several non-metals when heated, e.g.

$$Mg(s) + Cl_2(g) \longrightarrow MgCl_2(s)$$

$$Mg(s) + S(s) \longrightarrow MgS(s)$$

Compounds of magnesium

Magnesium oxide, MgO, is obtained by heating the nitrate or carbonate. White powder, slightly soluble in water forming hydroxide. Basic oxide forming salts with acids. Because of its high melting point, it is used as a refractory lining. Oxide also used to combat excess stomach acidity as magnesia.

Magnesium hydride, MgH_2, is obtained by heating magnesium powder in hydrogen under pressure with magnesium iodide as catalyst. Reacts with water to form hydrogen but is decomposed by heat.

$$MgH_2(s) + 2H_2O(l) \longrightarrow Mg(OH)_2(s) + 2H_2(g)$$

Magnesium carbonate, $MgCO_3$, is precipitated by sodium hydrogencarbonate, which is less hydrolysed than sodium carbonate, the concentration of hydroxyl ions is insignificant.

$$M^{2+}(aq) + 2HCO_3^-(aq) + 2H_2O(l) \longrightarrow MgCO_3.2H_2O(s) + CO_2(g)$$

Sodium carbonate precipitates basic carbonates. Magnesium carbonate is not precipitated with the carbonates of calcium, strontium and barium in analysis because $NH_4^+(aq)$ from the added ammonium carbonate and $Cl^-(aq)$ form hydrochloric acid. Carbonate forms oxide on heating.

Magnesium chloride, $MgCl_2$.   The action of dilute hydrochloric acid on the metal, oxide or carbonate yields $MgCl_2.6H_2O$ on crystallization. On heating,

$$MgCl_2.6H_2O \rightleftharpoons Mg(OH)_2Cl(s) + HCl(g) + 5H_2O(l)$$

On heating more strongly,

$$Mg(OH)Cl(s) \longrightarrow MgO(s) + HCl(g)$$

Anhydrous magnesium chloride is obtained by heating magnesium in dry chlorine or dry hydrogen chloride. Equimolar masses of ammonium chloride and magnesium chloride form a double salt.

Magnesium sulphate, $MgSO_4.7H_2O$, Epsom salts. Obtained by the action of dilute sulphuric acid on the metal, oxide or carbonate. (See previous note on hydration.) Epsom salts are used in medicine.

Similarities between magnesium and zinc
1 Fairly low melting points.
2 Hydrides are covalent.
3 React slowly with water.
4 Low affinity for oxygen and chlorine.
5 Sulphates soluble in water and highly hydrated.
6 Salts less stable to heat.

Calcium, Ca

Occurrence   **Calcium carbonate**, $CaCO_3$, as chalk, marble, calcspar, Iceland spar, dolomite $CaCO_3.MgCO_3$; **calcium sulphate**, $CaSO_4$, anhydrite, also gypsum $CaSO_4.2H_2O$; **fluorspar** CaF_2; phosphates in bones.

Extraction   **Electrolysis** of fused **calcium chloride** mixed with one-sixth of its weight of calcium fluoride to lower the temperature. The **cathode** is a **water-cooled steel tube** and the **anode is graphite**. Solid calcium sticks to the bottom of the cathode which is raised as the metal accumulates so that calcium soon becomes the cathode. A layer of electrolyte protects the metal from oxidation. The calcium is freed from electrolyte by melting it in argon and it is then purified by distillation *in vacuo*.

Uses
1 Deoxidant for metals, e.g. copper, aluminium and some steels.
2 Hardening lead in alloys.
3 Manufacture of hydrolith.
4 Important biologically.

Physical properties
Silvery metal that can be cut with a knife.

Chemical properties
1 On exposure to air, forms the oxide, then hydroxide and then carbonate.
2 When heated gently in air, forms a mixture of oxide and nitride.
3 Nitride forms ammonia with water.

$$Ca_3N_2(s) + 6H_2O(l) \longrightarrow 3Ca(OH)_2(s) + 2NH_3(g)$$

4 Reacts readily with water and sinks (alkali metals float). Reaction is slowed down by the insolubility of calcium hydroxide.

$$Ca(s) + 2H^+(aq) + 2OH^-(aq) \longrightarrow Ca(OH)_2(s) + H_2(g)$$

5 Combines when heated in dry hydrogen.
6 Does not react with alkalis.

Compounds of calcium

Calcium hydride, $Ca^{2+}(H^-)_2$ hydrolith. Salt-like hydride formed by heating calcium in dry hydrogen. Forms hydrogen with water.

$$CaH_2(s) + 2H_2O(l) \longrightarrow Ca(OH)_2(s) + 2H_2(g)$$

Quicklime, CaO, is obtained by heating calcium carbonate in a lime kiln. At 1170 K decomposition is almost complete.

$$CaCO_3(s) \longrightarrow CaO(s) + CO_2(g)$$

Properties
1 White, amorphous solid.
2 Highly refractory.
3 Becomes incandescent when very hot – limelight.
4 Reacts vigorously and exothermically with water giving slaked lime.

$$CaO(s) + H_2O(l) \longrightarrow Ca(OH)_2(s)$$

5 Combines directly with chlorine at red-heat forming calcium chloride.

$$2CaO(s) + 2Cl_2(g) \longrightarrow CaCl_2 + O_2(g)$$

Uses
1 Used in mortar and plaster and in the manufacturing of bleaching powder.
2 Soda lime is formed by slaking quicklime with sodium hydroxide solution.
3 Used as a drying agent for ethanol and ammonia gas.

Calcium hydroxide, $Ca^{2+}(OH^-)_2$, slaked lime.

Properties
1 White solid.
2 Solubility decreases as temperature rises. A saturated solution is lime water, used as a test for carbon dioxide: clear solution becomes cloudy as calcium carbonate is precipitated.

$$Ca(OH)_2(aq) + CO_2(g) \longrightarrow CaCO_3(s) + H_2O(l)$$

3 Reacts with acids to form salts. Reaction with dilute sulphuric acid stops because calcium sulphate is insoluble.
4 Frees ammonia from ammonium salts on heating.

$$NH_4^+(s) + OH^-(s) \longrightarrow NH_3(g) + H_2O(l)$$

5 With chlorine, at ordinary temperature, forms bleaching powder.

$$Ca(OH)_2(s) + Cl_2(g) \longrightarrow CaOCl_2 + H_2O(l)$$

At red-heat calcium chloride and oxygen are formed

$$2Ca(OH)_2(s) + 2Cl_2(g) \longrightarrow 2CaCl_2(s) + 2H_2O(l) + O_2(g)$$

Cold milk of lime forms a mixture of calcium chloride and calcium hypochlorite.

$$Ca(OH)_2(s) + Cl_2(g) + H_2O(l) \longrightarrow$$
$$Ca^{2+}(aq) + Cl^-(aq) + OCl^-(aq) + 2H_2O(l)$$

Hot milk of lime yields calcium chloride and calcium chlorate(V).

$$3Ca(OH)_2(s) + 3H_2O(l) + 3Cl_2(g) \longrightarrow$$
$$3Ca^{2+}(aq) + 5Cl^-(aq) + ClO_3^-(aq) + 6H_2O(l)$$

Uses
1 Softening of temporary hard water.
2 Used in the production of mortar, plaster and bleaching powder.
3 Used in the Solvay process.

Calcium carbonate, $Ca^{2+}CO_3^{2-}$. Decomposes reversibly on heating.

$$CaCO_3(s) \rightleftharpoons CaO(s) + CO_2(g)$$

Calcium hydrogencarbonate, $Ca^{2+}(HCO_3^-)_2$. Formed when carbon dioxide is passed into lime water. Cause of temporary hard water and may be removed by boiling.

Calcium dicarbide, $Ca^{2+}(C\equiv C)^{2-}$. Formed when an electric arc is struck between carbon electrodes in a furnace containing coke and quicklime.

$$CaO(s) + 3C(s) \longrightarrow CaC_2(s) + CO(g)$$

1 Forms ethyne with water

$$CaC_2(s) + 2H_2O(l) \longrightarrow Ca(OH)_2(s) + C_2H_2(g)$$

2 Forms calcium cyanamide with nitrogen and used as a nitrogenous fertilizer -- nitrolin.

$$CaC_2(s) + N_2(g) \longrightarrow CaCN_2(s) + C(s)$$

Phosphates of calcium Used as fertilizers.

Barium, Ba

Occurrence As **barytes** or heavy spar, $BaSO_4$, or **witherite**, $BaCO_3$.

Extraction **Thermit** process used because metal is very reactive.

Barium oxide is heated with **aluminium** powder at white heat (1500 K) at very low pressure.

$$3BaO(s) + 2Al(s) \longrightarrow Al_2O_3(s) + 3Ba(s)$$

Properties
Soft silvery metal, resembles strontium and calcium, but more reactive.

Compounds of barium

Barium oxide, $Ba^{2+}O^{2-}$. Obtained in the laboratory by heating barium nitrate.

$$2Ba(NO_3)_2(s) \longrightarrow 2BaO(s) + 4NO_2(g) + O_2(g)$$

Slaked by water to form barium hydroxide. Basic oxide but does not react readily with dilute sulphuric acid because barium sulphate is insoluble.

Barium peroxide, BaO_2. Formed by heating the oxide in oxygen. Reaction is reversible and was the basis of the Brin process for extracting oxygen from the air. Crystallizes as $BaO_2.8H_2O$ if hydrogen peroxide is added to a cold, saturated solution of barium hydroxide.

Barium hydroxide, $Ba^{2+}(OH^-)_2$. More soluble than the hydroxides of calcium and strontium and can be used to titrate weak acids – baryta water. Is not contaminated with barium carbonate because the latter is so insoluble.

Barium chloride crystallizes as $BaCl_2.2H_2O$ and differs from $CaCl_2.6H_2O$ and $SrCl_2.6H_2O$ in that it is not deliquescent.

Barium sulphate, $BaSO_4$, is very insoluble but is difficult to obtain pure because it absorbs other salts, e.g. barium chloride. It is used in paints and because it is relatively opaque to X-rays, it is given as a 'barium meal' in medical examinations.

Barium sulphide, BaS, is obtained by the reduction of barium sulphate with coke.

$$BaSO_4(s) + 4C(s) \longrightarrow BaS(s) + 4CO(g)$$

It is soluble and is hydrolysed in water.

$$S^{2-}(s) + 2H_2O(l) \rightleftharpoons 2OH^-(aq) + 2SH^-(aq)$$

Impure barium sulphide is phosphorescent.

Barium nitrate, $Ba(NO_3)_2$, is less soluble than most nitrates. Used to produce green in fireworks.

Practice questions

1 Write out the electronic configurations of $_4Be$, $_{20}Ca$, $_{20}Ca^{2+}$, $_{56}Ba$, $_{56}Ba^{2+}$.

2 Explain why the atoms of the alkaline earth metals are smaller than those of the alkali metals.

3 Suggest why the alkaline earth metals form ions M^{2+} although the ionization energies of the atoms are high.

4 Outline five ways in which the alkaline earth metals differ from the alkali metals.

5 Outline the differences down group II in the ways in which the metals react.

6 Comment on the degree of hydration of   (a) magnesium sulphate,   (b) calcium sulphate.

7 In what ways does beryllium differ from the other alkaline earth metals? Suggest a reason for these differences.

8 How is magnesium extracted?

9 How does magnesium react with   (a) air,   (b) water,   (c) hot concentrated sulphuric acid,   (d) sulphur?

10 How is magnesium carbonate obtained? Explain why you have selected the reagents in the reaction.

11 How is anhydrous magnesium chloride prepared?

12 How is calcium extracted?

13 How does   (a) calcium hydride,   (b) calcium nitride react with water?

14 How does calcium dicarbide react with
(a) water,   (b) nitrogen?

15 Write equations and give the experimental conditions necessary for preparing   (a) NaOH,   (b) Na_2CO_3,   (c) $NaHCO_3$,
(d) KCl,   (e) CaO,   (f) $Ca(HCO_3)_2$,   (g) $MgCl_2$,
(h) $MgSO_4$.

16 A solution contains Na^+, Be^{2+} and Mg^{2+}. How would you separate the ions from each other?

17 Give two tests in each case (six tests in all) which would enable you to distinguish and then confirm the compounds in each of the following mixtures
(a) sodium carbonate and potassium carbonate,
(b) calcium oxide and calcium carbonate,
(c) barium chloride and barium nitrate.

7   p-BLOCK ELEMENTS: METALS OF GROUP III

Boron, aluminium, gallium, indium and thallium all have an outer electronic configuration of ns^2p^1 and oxidation states of +1 or +3.

<table>
<tr><td rowspan="5">Oxidation states</td><td>3 B</td><td colspan="2">$1s^2$ $2s^2$ $2p^1$</td></tr>
</table>

3 B	$1s^2$	$2s^2$ $2p^1$					
1, 3 Al	$1s^2$	$2s^2$ $2p^6$	$3s^2$ $3p^1$				
3 Ga	$1s^2$	$2s^2$ $2p^6$	$3s^2$ $3p^6$	$3d^{10}$ $4s^2$ $4p^1$			
1, 3 In	krypton core	$4d^{10}$		$5s^2$ $5p^1$			
1, 3 Tl	krypton core	$4d^{10}$ $4f^{14}$		$5s^2$ $5p^6$	$5d^{10}$	$6s^2$ $6p^1$	

General properties

1 Not strongly electropositive.

2 High ionization energy for group III metals means that they do not form ionic compounds like the elements of groups I and II and most form covalent compounds.

3 3-covalent compounds are usually trigonal planar in shape.

4 Electronegativity increases down the group. The outer electrons of indium and thallium are held more tightly because of the poor screening effect of the d-electrons.

5 Gallium, indium and thallium have d-electrons and higher densities than boron and aluminium.

6 After aluminium, the tendency for the s-electrons to remain paired, the inert pair effect, increases so that for thallium, the oxidation state +1 is more stable than +3.

7 Ionization energy decreases down the group so that boron forms covalent compounds and aluminium forms hydrated salts, e.g. $[Al(H_2O)_6]^{3+}$ is known in solutions of aluminium salts and in solid hydrated salts, $AlCl_3.6H_2O$. Compound AlF_3 may be partly covalent.

8 Boron hydroxide is weakly acidic, aluminium hydroxide is amphoteric and the remaining hydroxides are basic.

9 Form complexes which may be 4 co-ordinate and tetrahedral, e.g. $[BF_4]^{2-}$ and 6 co-ordinate, octahedral, e.g. $[AlF_6]^{3-}$.

10 Complexes may be partly ionic and partly covalent. If covalent, electrons which are donated to the metal may enter the third shell in aluminium which can hold up to eighteen electrons, e.g. $[AlF_6]^{3-}$. In boron the second shell is available for accepting electrons but this is filled with eight electrons so a maximum of 4-covalent bonds is possible as in $[BF_4]^-$. This increase in the maximum possible covalency between first and second elements is characteristic of groups III–VII but the maximum covalency is only exerted in the presence of strongly electronegative atoms or groups. Thus, while there are $[AlF_6]^{3-}$, $[SiF_6]^{2-}$, SF_6, there are no corresponding stable chloride compounds.

Boron, B

Occurrence   **Trioxoboric acid**, boric acid, H_3BO_3 and as borates.

Properties

When pure, boron is a very hard black powder with a metallic appearance but it is a very poor conductor of electricity. Amorphous boron is a brown powder. It combines directly with nitrogen and oxygen forming BN and B_2O_3. It combines with fluorine in the cold forming BF_3 and on heating combines with chlorine and sulphur to form BCl_3 and B_2S_3. It reacts only with oxidizing acids forming trioxoboric acid

$$2B(s) + 6HNO_3(conc.aq.) \longrightarrow 2H_3BO_3(s) + 6NO_2(g)$$

Forms a borate, trioxoborate(III) with sodium or potassium hydroxide.

$$2B(s) + 6KOH(s) \longrightarrow 2K_3BO_3(l) + 3H_2(g)$$

When strongly heated, boron reduces silica and carbon dioxide.

$$4B(s) + 3SiO_2(s) \longrightarrow 3Si(s) + 2B_2O_3(s)$$

Uses

1   As a deoxidizer in steel, when used in small amounts.
2   In larger amounts, it forms boron steels which are very strong.

Compounds of boron

Diboron trioxide, B_2O_3, is a glassy, hygroscopic solid which is amphoteric.

Boron hydridres. Fourteen are known, e.g. B_2H_6, B_4H_{10}, B_5H_9, B_5H_{11}, B_6H_{10}, B_6H_{12} and $B_{10}H_{14}$.

Diborane is not known as BH_3 but as B_2H_6. It is obtained by the reduction of an ethereal solution of boron trichloride by lithium aluminium hydride. It contains electron-deficient bonds or hydrogen-bridge structures. The orbitals linking the boron atoms are sp^3 hybrid orbitals and are banana-shaped and differ from the orbitals in, e.g. ethene, C_2H_6.

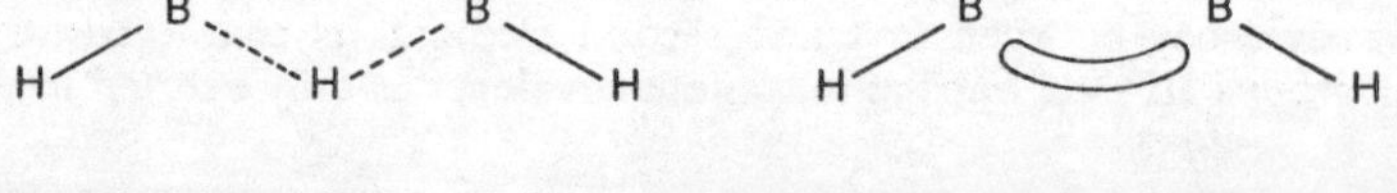

Borazine, $B_3N_3H_6$.   Ammonia reacts with diborane to form a structure with similar physical properties to benzene but the B—N bonds are less stable than C—H bonds.

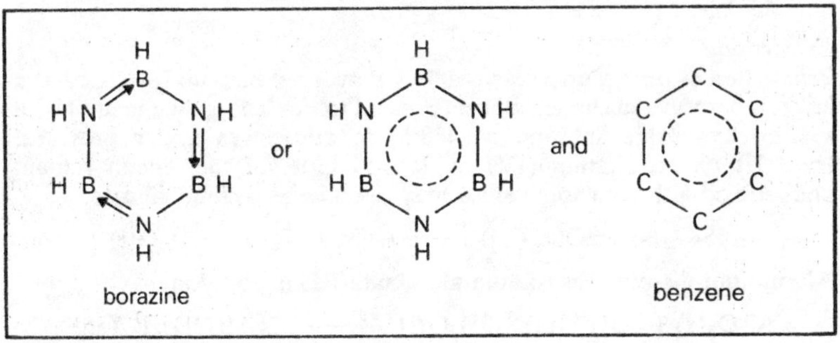

Boron trifluoride, BF_3, has only six electrons in the outer shell and can accept a lone pair. It can form an addition compound with ammonia, $H_3N.BF_3$. Here BF_3 is acting as the acceptor or Lewis acid and ammonia is the donor or Lewis base. BCl_3 can also act as a Lewis acid towards ammonia. Boron trichloride is hydrolysed,

$$BX_3 \underset{}{\overset{H_2O}{\rightleftharpoons}} BX_2OH \underset{}{\overset{H_2O}{\rightleftharpoons}} BX(OH)_2 \underset{}{\overset{H_2O}{\rightleftharpoons}} B(OH)_3$$
$$+ \qquad\qquad + \qquad\qquad +$$
$$HCl \qquad\qquad HCl \qquad\qquad HCl$$

This substitution does not occur with the fluoride because of the strength of the B—F bond but it forms 1/1 and 1/2 adducts with water.

$$BF_3(g) + H_2O(l) \rightleftharpoons H^+[BF_3OH]^-(aq) \overset{H_2O}{\rightleftharpoons}$$
$$H_3O^+[BF_3OH^-](aq)$$

Metal borides such as CaB_6, AlB_2, TiB_2 and FeB formed by heating the elements in powdered form at a high temperature in a vacuum have structures like those of interstitial carbides and the precise formulae are difficult to determine.

Uses of boron compounds
1 Boric acid or boracic acid is used as a mild antiseptic, e.g. in eye lotions and also as a glaze for enamels on metallic surfaces.
2 Borax is used in the borax bead test, as a flux in welding, in making low expansion glass, e.g. Pyrex and in glazing card and paper.
3 Sodium perborate is used as an oxidizing agent and as an antiseptic, used in washing powders.

Aluminium, Al

Occurrence The most abundant metal. Found as the **oxide** Al_2O_3, corundum; **hydrated oxide** $Al_2O_3.2H_2O$, e.g. bauxite; fluoride Na_3AlF_6, cryolite and as **silicates**. Some gems are transparent forms of aluminium oxide, e.g. sapphire is blue due to cobalt or titanium.

Extraction

Production of pure aluminium oxide. Powdered bauxite is roasted at a low temperature to change all iron present to Fe^{3+}. It is then heated with **sodium hydroxide** solution at 430 K in autoclaves under pressure. Iron(III) oxide, titanium(IV) oxide and most of the silicon remain undissolved although some silica may dissolve as sodium silicate.

$$SiO_2(s) + 2OH^-(aq) \longrightarrow SiO_3{}^{2-}(aq) + H_2O(l)$$

Aluminium dissolves as **sodium aluminate(III)** in solution.

$$Al_2O_3(s) + 6OH^-(aq) + 3H_2O(l) \longrightarrow 2[Al(OH)_6]^{3-}(aq)$$

Undissolved precipitates are filtered off and the liquid left is diluted and 'seeded' with a little freshly prepared aluminium hydroxide and stirred. **Aluminium hydroxide** is precipitated and any silicates remain in solution.

$$[Al(OH)_6]^{3-}(aq) \longrightarrow Al(OH)_3(s) + 3OH^-(aq)$$

The hydroxide is filtered, washed, dried and heated to form pure aluminium oxide which is electrolysed.

Electrolysis of aluminium oxide

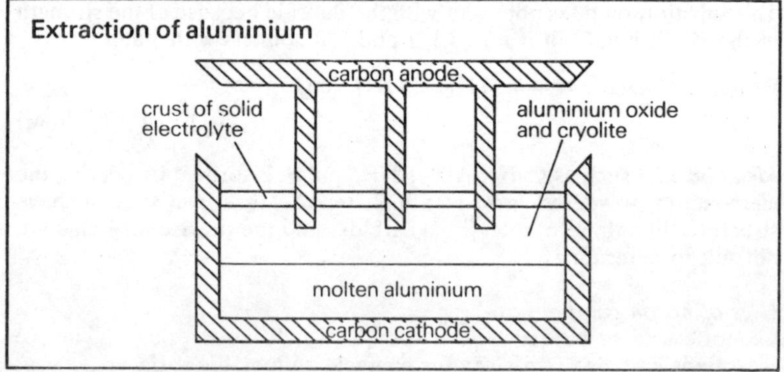

The **aluminium oxide** is dissolved in a molten mixture of **cryolite**, Na_3AlF_6, and **fluorspar**, CaF_2. Providing the concentration of aluminium oxide is maintained at a sufficiently high level, the fluorides are not affected. The steel cell is lined with carbon and the anodes consist of carbon rods dipping into the melt. Aluminium collects on the floor of the cell and oxygen formed at the anode burns away the carbon rods.

Cathode

$$2Al^{3+} + 6e^- \longrightarrow 2Al(l)$$

Anode

$$3O^{2-} - 6e^- \longrightarrow 1\tfrac{1}{2}O_2(g)$$

Reduction of aluminium ion

The aluminium obtained is 99% pure. Because of its high valency and low relative atomic mass, aluminium is expensive to produce. Hydro-electricity is used.

Uses
1 Construction of doors and window frames – does not corrode or warp.
2 In cooking utensils – bright, light and easy to clean but is attacked by alkalis.
3 Overhead high tension cables. Conducts well and is light.
4 In light alloys, e.g. magnalium and duralumin. Most important in aircraft construction.
5 Thermite process to reduce oxides of other metals. Aluminium powder used in welding steel.
6 As a packing material.
7 As a paint.
8 As a protective metal. When in contact with the atmosphere, aluminium acquires a thin layer of oxide which adheres and protects it against further attack.
9 **Anodizing** is the electrolytic oxidation of aluminium making the layer of oxide thicker. The metal is made the anode, the cathode is steel and the electrolyte a 3% solution of chromic(VI) acid. The oxide layer is porous at first and can absorb coloured dyes.

Physical properties
Bluish-white metal that can be highly polished but is usually dulled by a layer of oxide. Conducts heat and electricity well, is malleable and ductile, used to make foil and wires.

Chemical properties
1 Burns in air to form oxide and nitride. Nitride forms ammonia with water.

$$4Al(s) + 3O_2(s) \longrightarrow 2Al_2O_3(s)$$

$$2Al(s) + N_2(g) \longrightarrow 2AlN(s)$$

$$AlN(s) + 3H_2O \longrightarrow Al(OH)_3(s) + NH_3(g)$$

Metal burns readily if heated and was once used in flashlight photography. If aluminium is dipped into mercury(II) solution an

Al/Hg alloy is formed and the film of oxide cannot cohere. The activity of aluminium can then be seen and the metal oxidizes rapidly. Spillage of mercury can destroy aluminium instruments.

2 Reacts with dilute hydrochloric acid to form aluminium chloride and hydrogen. No reaction with dilute sulphuric acid is possibly because the oxide layer is insoluble in the acid. Hot concentrated sulphuric acid forms sulphate and sulphur dioxide.

$$2Al(s) + 6H_2SO_4(aq) \longrightarrow Al_2(SO_4)_3(aq) + 6H_2O(l) + 3SO_2(g)$$

Nitric acid has very little effect even when hot and concentrated.

3 Reacts with cold dilute alkali – powder readily, foil on warming, to form aluminates.

$$2Al(s) + 2OH^-(aq) + 10H_2O(l) \longrightarrow$$
$$2[Al(OH)_4(H_2O)_2]^-(aq) + 3H_2(g)$$

4 Combines when heated with chlorine, nitrogen, sulphur and carbon forming $AlCl_3$, AlN, Al_2S_3 and Al_4C_3 respectively.

Compounds of aluminium

Aluminium hydroxide, $Al^{3+}(OH)_3$, is deposited as a gelatinous precipitate when ammonia solution is added to a solution of an aluminium salt.

$$Al^{3+}(aq) + 3OH^-(aq) \longrightarrow Al(OH)_3(s)$$

If the mixture is heated, the hydroxide becomes amorphous and can be filtered and dried. The gelatinous precipitate contains water, $2H_2O$, and is octahedral, $[Al(H_2O)_2OH_4]^-$.

The aluminium ion, Al^{3+}, is so small and highly charged that it is heavily hydrated. The complex cation functions as a triprotic acid releasing protons to form oxonium ions with water molecules. The acid strength is such that it can be titrated against sodium hydroxide.

$$[Al(H_2O)_6]^{3+} \quad [Al(OH)_3(H_2O)_3] \quad [Al(OH)_6]^{3-}$$

$$\xrightarrow{\quad\text{alkali}\quad}$$
$$\xleftarrow{\quad\text{acid}\quad}$$

Salts of weak acids, e.g. carbonates and sulphides, cannot be prepared in solution because the anions behave as strong bases precipitating the hydroxides.

$$2[Al(H_2O)_6]^{3+}(aq) + 3CO_3{}^{2-}(aq) \longrightarrow$$
$$2[Al(OH)_3H_2O](s) + 3CO_2(g) + 3H_2O(l)$$

Addition of excess alkali forms $[Al(OH)_6]^{3-}$ which dissolves as aluminate.

Uses
1 As a mordant in dyeing.
2 Treatment of sewage.

Aluminium oxide, Al_2O_3. Highly refractive white powder, melting point 2300 K. Amphoteric, but if heated above 1200 K becomes denser and is almost unaffected by acids. Similar properties are shown by Cr_2O_3, Fe_2O_3 and MgO.

Uses
1 As a source of aluminium.
2 In cement.
3 As a refractory substance.
4 In abrasives.
5 In chromatography.

Aluminium chloride, $AlCl_3$, obtained by heating aluminium in dry chlorine. White when pure but often yellow due to the presence of iron(III) chloride. Fumes in air because of hydrolysis and the formation of HCl. Sublimes at 456 K and the vapour density corresponds to Al_2Cl_6. Above 620 K dissociation occurs.

$$Al_2Cl_6 \rightleftharpoons 2AlCl_3$$

Al_2Cl_6 is covalently bonded and the chlorine atoms can be considered as donating a lone pair of electrons to the aluminium atom in the other 'half' of the molecule so that the outer shell of each aluminium atom contains 8 electrons.

The shape is trigonal planar like that of BCl_3^-.

Uses

1   As a catalyst, e.g. in Freidel Crafts reaction.

$$C_6H_6 + RCl \longrightarrow C_6H_5R + HCl$$

Aluminium sulphate, $Al_2(SO_4)_3 . 18H_2O$. Made by the action of moderately concentrated sulphuric acid on bauxite.

Uses

1   Purification of water.
2   Tanning leather.
3   Sizing of paper.
4   In foam fire extinguishers.

Alums, double salts $X^+Y^{3+}(SO_4^{2-})_2 . 12H_2O$, where X = Na, K or NH_4 and Y = Al, Cr(III) or Fe(II).   Isomorphous crystallizing occurs in the same octahedral form and mixed crystals can be obtained. Can be prepared by mixing equimolar amounts of the two metal sulphates to the minimum volume of hot water that will dissolve them, mixing well and cooling until the alum crystallizes.

Similarities between aluminium and beryllium

1   Melting points higher than expected.
2   React with sodium hydroxide.
3   Less electropositive than expected and form covalent compounds.
4   Hydrides are difficult to prepare.
5   Salts are hydrolysed.

Practice questions

1   Explain the type of valency shown by the elements of group III.

2   Write brief notes on complex formation in group III.

3   Suggest electronic structures for   (a) diborane,   (b) borazine.

4   Boron trichloride is hydrolysed by water. How does the action of boron trifluoride on water differ from this?

5   Name two ores of aluminium and outline the way in which aluminium is extracted from one of them.

6   When the surface of a piece of aluminium foil is scratched with mercury(II) chloride, the aluminium is oxidized rapidly. Comment on this.

7   What is   (a) a sacrificial metal,   (b) anodizing?

8   Explain why aluminium carbonate cannot be prepared in solution.

9 Suggest a formula and electronic structure for aluminium chloride.

10 What are alums? How would you obtain dry crystals of potash alum? In your answer explain the meaning of isomorphous.

11 Outline the similarities between aluminium and beryllium.

8 p-BLOCK ELEMENTS OF GROUP IV

Carbon, silicon, germanium, tin and lead all have an outer electronic configuration of $ns^2\, np^2$ and have oxidation states of +2 and +4.

<table>
<tr><td rowspan="5" style="writing-mode: vertical-rl">Oxidation states</td></tr>
<tr><td>2, 4 C</td><td>$1s^2\ 2s^2\ 2p^2$</td></tr>
<tr><td>2, 4 Si</td><td>$1s^2\ 2s^2\ 2p^6\ 3s^2\ 3p^2$</td></tr>
<tr><td>2, 4 Ge</td><td>$1s^2\ 2s^2\ 2p^6\ 3s^2\ 3p^6\ 3d^{10}\ 4s^2\ 4p^1$</td></tr>
<tr><td>2, 4 Sn</td><td>krypton core $\quad 4d^{10}\qquad 5s^2\ 5p^2$</td></tr>
<tr><td>2, 4 Pb</td><td>$4d^{10}\ 4f^{14}\ 5s^2\ 5p^6\ 5d^{10}\ 6s^2\ 6p^2$</td></tr>
</table>

General properties

1 Elements have little tendency to form M^{4+} because of the high values for the third and fourth ionization energies.

2 The bonding in the tetravalent compounds is usually covalent, the bonds obtained by the hydridization of the sp-orbitals.

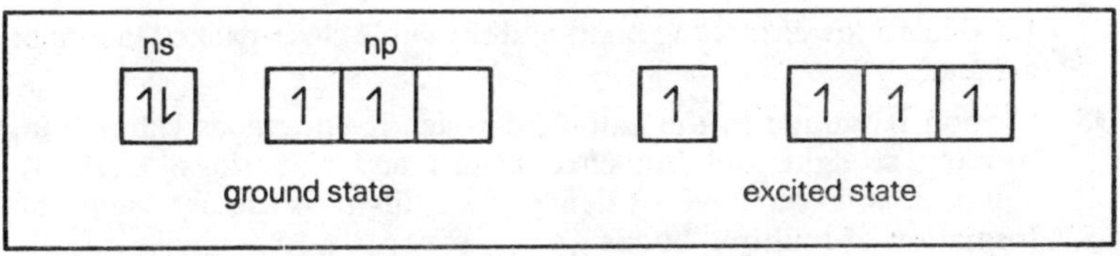

3 Carbon and silicon are almost always 4-covalent, the four bonds pointing towards the corners of a tetrahedron.

4 The tendency to form divalent compounds increases down the group, the s-electrons tending to remain paired (inert pair effect) in the heavier elements.

5 Tin(IV) compounds are more stable than tin(II) compounds so that tin(II) compounds are reducing agents.

$$Sn^{2+} \longrightarrow Sn^{4+} + 2e^-$$

6 Lead(II) compounds are more stable than lead(IV) compounds so that lead(IV) compounds are covalent and are strong oxidizing agents.

$$Pb^{4+} + 2e^- \longrightarrow Pb^{2+}$$

7   As the stability of the lower oxidation state increases down the group the tendency to form ionic bonds also increases. Thus, the elements at the bottom of the group form M^{2+} ions more readily than the elements at the top. Ge^{2+} is a strong reducing agent but Sn^{2+} and Pb^{2+} are found in the solid state and in solution.

8   The number of possible bonds that can be formed by an element increases sharply after carbon because of the presence of d-orbitals. Carbon is limited to four covalent bonds because it has no d-orbitals but the tetrafluorides of silicon to lead dissolve in excess hydrogen fluoride forming $[MF_6]^{2-}$. In excess hydrochloric acid, tin(IV) chloride and lead(IV) chloride form $[SnCl_6]^{2-}$ and $[PbCl_6]^{2-}$.

9   A changeover from non-metallic to metallic character occurs down the group and the elements become more electropositive.

10   Carbon and silicon are non-metals, germanium shows some metallic characteristics while tin and lead are metals.

11   The oxides of carbon and silicon are acidic or neutral; the oxides of tetravalent tin and lead are amphoteric.

12   Carbon exists as diamond and graphite. Silicon and germanium both have diamond-type lattices with high melting points. Tin has three polymorphic forms:

Grey (α) tin	$\rightleftharpoons$	White (β) tin	$\rightleftharpoons$	Brittle tin
diamond		body centre		rhombic
lattice		cubic structure		

Lead has a lower melting point and the cubic close-packed lattice of a metal.

13   Carbon is unique in the extent to which it undergoes catenation, forming straight and branched chains and also rings. Carbon's apparent ability to have a valency of less than 4 is usually due to the formation of multiple bonds.

14   Silicon differs from carbon in having a greater affinity for oxygen forming silicates and silicones.

15   Silicon does not form multiple bonds with itself. The C—C bond is much stronger than the Si—Si bond. The other elements in group IV do not show catenation.

16   Electronegativity varies as in group III and this is characteristic of elements containing inner d-electrons.

17   Carbon has much higher melting and boiling points than the other elements in its group; it sublimes as graphite at 4000 K and boils as graphite at 5100 K.

18   All the group form covalent hydrides, those formed by carbon being the basis of organic chemistry. Silicon and germanium form a number of hydrides of formulae Si_nH_{2n+2} and Ge_nH_{2n+2n}, where

$n = 7$ or 8. Tin forms stannane, SnH_4, and lead forms plumbane, PbH_4.

19 The stability and volatility of the hydrides decreases down the groups so that plumbane is unstable at room temperature. Methane is very stable and has weak reducing properties, e.g. can reduce copper(II) oxide. Tetrahydrides of the other elements are strong reducing agents. This is probably due to the differences in the polarity of the bonds in the different hydrides. In methane, the bond is polarized $C^{\delta-}$—$H^{\delta+}$. In the others, it is $M^{\delta+}$═$H^{\delta-}$, the degree of polarization and ionic character increasing down the group.

20 All hydrides can be obtained by the action of $LiAlH_4$ on the corresponding tetrahalide.

$$MX_4 + LiAlH_4 \xrightarrow{\text{dry ethoxyethane}} MH_4 + LiAlX_4$$

21 Oxides of carbon are gaseous and neutral, CO; or acidic CO_2, C_3O_2; while the oxides of silicon are solids. SiO is unstable and SiO_2 is acidic. Germanium oxides show the properties expected. GeO solid and amphoteric, GeO_2 solid and mainly acidic. The oxides of tin, SnO and SnO_2, are amphoteric solids while those of lead are basic with some amphoteric properties; PbO is amphoteric but mainly basic, PbO_2 is amphoteric and Pb_3O_4 is a 'mixed' oxide.

22 Halides are numerous because of carbon catenation. Chloro-fluorohydrocarbons (freons) are important as refrigerants and propellants in aerosols. Polytetrafluoroethane PTFE or Teflon is used in non-stick cooking utensils.

23 Silicon forms halides, $Si_n Cl_{2n+2}$, the longest being Si_6Cl_{14}.

24 Except for PbI_4 and perhaps $PbBr_4$, all MX_4 are known. PbI_4 cannot exist because of the strong reducing properties of I^- and the strong oxidizing properties of Pb^{4+}. The fluorides SnF_4 and PbF_4 are ionic but the other halides are covalent with tetrahedral structures.

25 The stability of tetravalent halides decreases from carbon to lead. The formation of four covalent bonds by carbon as in CCl_4, confers stability on the carbon atom and the molecule is not attacked by water. The silicon atom in $SiCl_4$ is susceptible to attack by water and it is hydrolysed to a hydrated oxide (silicic acid) formed as a colloidal solution or gel. Both $GeCl_2$ (a solid) and $GeCl_4$ (a liquid) are hydrolysed. $SnCl_2$ (a solid) shows some ionic characteristics and is slowly hydrolysed by water. $SnCl_4$ (a liquid) fumes in moist air and is soluble in organic solvents. $PbCl_2$ (a solid) is ionic and is insoluble in cold water but dissolves in hot water. $PbCl_4$ (a liquid) is hydrolysed by water but the hydrolysis of $SnCl_4$ and $PbCl_4$ is prevented by acids.

26 Carbides – the electronegativity of the group IV elements is too low to form anions although ionic carbides, $(C\equiv C)^{2-}$ are formed by group I and group II elements. They form ethyne, $HC\equiv CH$, on hydrolysis. Beryllium and aluminium carbides Be_2C and Al_4C_3 are partly covalent yielding methane on hydrolysis. Silicon(IV) carbide, SiC (carborundum) and boron(III) carbide, B_4C_3, are almost entirely covalent with giant structures, very hard, abrasive and chemically inert.

27 Carbon forms interstitial carbides with the transition metals. The carbon atoms occupy spaces in the octahedral lattice – they are inert and able to conduct electricity.

Carbon, C

Occurrence In many compounds e.g. hydrocarbons, carbohydrates, and as coal and petroleum compounds.

Allotropy – two allotropes, diamond and graphite.

Diamond is found mainly in the blue rock of South Africa. The round pebbles are cut and polished to make gems. Each diamond is a giant covalent structure with each carbon atom at the centre of a regular tetrahedron, the four valencies of each carbon atom directed towards the corners of the tetrahedron. All the bonding is covalent with no weaker van der Waal forces in the crystal. This uniformity of bonding makes carbon very hard. The structure is not broken down until over 3800 K when it vaporizes. It is transparent to X-rays while glass is opaque. Its high refractive index gives it its beauty. It is chemically unreactive, not attacked by acid or aqueous alkali. Reacts slowly with fused sodium carbonate.

$$C(s) + Na_2CO_3(l) \longrightarrow Na_2O(s) + 2CO(g)$$

A mixture of concentrated sulphuric acid and potassium dichromate(VI) at 470 K oxidizes diamond to carbon dioxide and fluorine changes it to CF_4 at red-heat. Much effort has been made to change inexpensive graphite to diamond but only a few commercial stones have been obtained.

Graphite Carbon atoms are covalently bonded in a network of regular hexagons forming parallel sheets held together by weak van der Waal forces. The sheets can shear or slide over each other giving graphite its greasy property. The bond lengths between the carbon atoms in the planes lie between that for $C-C$ and $C=C$ suggesting some delocalization of electrons – it is slightly longer than the bond in benzene in which the electrons are completely delocalized. The distance between the planes is too big for covalent bonding. The heat of combustion of graphite is less than for diamond and graphite is thermodynamically the more stable allotrope. It is mined as graphite, plumbago and black lead. It can be manufactured by passing an alternating current between carbon electrodes in a furnace containing a core of carbon, petroleum, coke containing some sand and coal-tar. The passage of the alternating current maintains a very high temperature for some 30 hours.

$$SiO_2 + 3C(s) \longrightarrow SiC(s) + 2CO(g)$$

$$SiC(s) \longrightarrow Si(g) + C(s)$$

Used in lead pencils, as a lubricant, as an electrical conductor, to make crucibles, in the cores of nuclear reactors and in aerospace where its high tensile strength and low density are useful. In perfect graphite, conduction should occur along the planes but not perpendicular to them. One of the valency electrons is free to move under the influence of an applied potential difference making graphite a good conductor. The more open structure of graphite makes it more chemically reactive than carbon but it is not attacked by chlorine, acids or alkalis. At red-heat, graphite burns with difficulty.

$$C(s) + O_2(g) \longrightarrow CO_2(g)$$

The reaction with nitric acid varies with the conditions.

Amorphous carbon is graphite in which the sheets are fragmented, with the planes at random. Prolonged action of fuming nitric acid changes it to mellitic acid which on distillation with lime forms benzene.

$$C_6(COOH)_6(s) + 6CaO(s) \longrightarrow C_6H_6(l) + 6CaCO_3$$

Amorphous carbon occurs as charcoal and soot. It is used in wood stoves, as a decolourizer, a deodorizer and to absorb gases. It is also used to take the yellow colour out of raw sugar.

Destructive distillation of wood   Wood heated in the absence of air yields wood gas, pyroligneous acid (water, ethanoic acid and methanol), wood tar and a residue of wood charcoal.

Coke is the residue left after the distillation of coal for coal gas. It is used as a fuel, in the production of producer gas, water gas and as an industrial reducing agent. It is more reactive than graphite. It is not affected by halogens although it combines with fluorine when heated forming CF_4. It burns in oxygen forming carbon dioxide and at 1300 K it is a powerful reducing agent.

$$PbO(s) + C(s) \longrightarrow Pb(l) + CO(g)$$

$$Fe_2O_3(s) + 3C(s) \longrightarrow 2Fe(l) + 3CO(g)$$

It reduces concentrated sulphuric and nitric acids when heated. At the temperatures found in the electric furnace, carbon combines with several elements forming carbides, e.g. CS_2, SiC, Al_2C_3, CaC_2.

Fuel gases
Natural gas is 90% methane and is found with petroleum.
Producer gas and water gas   Obtained by alternately blowing **air** and **steam** through a furnace full of **coke**. The passage of air yields a mixture of **carbon monoxide** and **nitrogen** or producer gas and the reaction is exothermic.

$$O_2(g) + 4N_2(g) + 2C(s) \longrightarrow 2CO(g) + 4N_2(g)$$
$$\text{producer gas}$$

When the temperature is sufficiently high, the air is switched off and steam is blown through the coke. A mixture of hydrogen and carbon monoxide or water gas is formed. The reaction is endothermic

$$H_2O(g) + C(s) \longrightarrow CO(g) + H_2(g)$$

When the temperature has dropped, the steam is replaced by the air blow. If the temperature falls below 1300 K, carbon dioxide is formed which is of no use as a fuel.

$$C(s) + 2H_2O(g) \longrightarrow CO_2(g) + 2H_2(g)$$

Uses
1   As a fuel and as an industrial source of hydrogen.

Coal gas   Mostly hydrogen and methane, formed by the bacterial decay, at high pressure and temperature, of vegetative matter. Over a very long period of time there is a gradual increase of hardness and carbon content.

Peat $\longrightarrow$ lignite coals $\longrightarrow$ bituminous coals $\longrightarrow$

steam coal $\longrightarrow$ anthracite

Compounds of carbon

Oxides of carbon

CO, CO_2, C_3O_2, C_5O_2, $C_{12}O_9$

Carbon monoxide, CO
Formed by the incomplete combination of hydrocarbons.

Preparation
1   The action of cold, concentrated sulphuric acid on sodium methanoate or hot concentrated sulphuric acid on ethanedioic acid.

$$HCOONa(s) + H_2SO_4(l) \longrightarrow NaHSO_4(s) + H_2O(l) + CO(g)$$
$$H_2C_2O_4(s) \xrightarrow[\text{heat}]{\text{conc } H_2SO_4} CO(g) + CO_2(g) + H_2O(l)$$

The gas is passed through potassium hydroxide solution to remove carbon dioxide and then collected over water or by upward delivery because it is less dense than air.

2   Pass carbon dioxide over charcoal heated in a combustion tube.

$$CO_2(g) + C(s) \longrightarrow 2CO(g)$$

Gas passed through potassium hydroxide solution to remove unchanged carbon dioxide.

Commercial production

During the production of water gas and producer gas.

Physical properties

Colourless, odourless gas, almost insoluble in water and difficult to liquefy.

1   Combines firmly with the haemoglobin of the blood and prevents it from combining with oxygen so that anoxia results. Very poisonous and dangerous gas.

2   Dangerous **after damp** formed by the partial oxidation of **fire damp** or methane.

$$2CH_4(g) + 3O_2(g) \longrightarrow 2CO(g) + 4H_2O(l)$$

The exhaust fumes of internal combustion engines contain carbon monoxide and a car should never be run with the garage doors closed. Similarly flues and chimneys should be swept regularly.

Chemical properties

1   Functions as the anhydride of methanoic acid but the gas is too insoluble in water to behave as an acidic oxide. When carbon monoxide is passed over sodium hydroxide at 470 K under pressure, sodium methanoate is produced.

$$CO(g) + NaOH(l) \longrightarrow HCOO^-Na^+(l)$$

2   A reducing agent which reduces metallic oxides, e.g. copper(II) oxide and iron(III) oxide.

$$CuO(s) + CO(g) \longrightarrow Cu(s) + CO_2(g)$$

It does not reduce aluminium and magnesium oxides.
Iodine(V) oxide is used to estimate carbon monoxide because the iodine produced can be determined.

$$I_2O_5(s) + 5CO(g) \longrightarrow I_2(s) + 5CO_2(g)$$

3   **Combustion**   Burns with a blue flame forming carbon dioxide – seen above coal fires.

4   **With metals** the lone pair of electrons on the carbon atom enables

carbon monoxide to function as a **ligand**, combining with transition metals to form complex carbonyls.

$$Ni(s) + 4CO(g) \rightleftharpoons Ni(CO)_4(g)$$

In nickel carbonyl, the nickel atom effectively controls 36 electrons. The molecule is tetrahedral with nickel at the centre.

5 In bright sunlight, carbon monoxide reduces chlorine and bromine forming the highly poisonous carbonyl halides.

$$CO(g) + Cl_2(g) \longrightarrow COCl_2(g)$$

Carbon oxide dichloride (phosgene), $COCl_2$, was used as a World War I gas.

6 A mixture of carbon monoxide and sulphur vapour, when passed through a heated tube, forms $COS(g)$.

Uses

1 As a fuel and a reducing agent, e.g. in the blast furnace.
2 As a fuel in producer gas and water gas.
3 In the purification of nickel.
4 Manufacture of carbon oxide dichloride.
5 Manufacture of methanol.

Structure

Carbon monoxide is isolectronic with nitrogen and probably contains a triple bond, $C{\equiv}O$ as in $N{\equiv}N$. Electron diffraction measurements confirm this.

Carbon dioxide, CO_2

Preparation

The action of dilute acid on a carbonate, usually dilute hydrochloric acid on marble chips in the cold. It is collected over warm water because it is fairly soluble, or by downward delivery because it is heavier than air.

$$CaCO_3(s) + 2HCl(aq) \longrightarrow CaCl_2(aq) + H_2O(l) + CO_2(g)$$

The salt formed must be soluble for the reaction to occur so that, e.g. calcium carbonate and dilute sulphuric acid cannot be used – calcium sulphate is insoluble.

Commercial production

1 Action of heat on limestone in a limekiln.

$$CaCO_3(s) \rightleftharpoons CaO(s) + CO_2(g)$$

2 Formed during brewing by fermentation.

$$C_6H_{12}O_6(aq) \xrightarrow[\text{zymase}]{\text{enzyme}} 2C_2H_5OH(aq) + 2CO_2(g)$$

Physical properties

1   Colourless, odourless gas, sharp taste, sparingly soluble in water.
2   Solid carbon dioxide sublimes at atmospheric pressure. Gas is stored as a solid, dry ice (Drikold) or as a liquid.

Chemical properties

1   The atmosphere contains approximately 0·03% and it is vital for life on Earth.

2   Sparingly soluble in water, the solution turns litmus dark red and a weak dibasic acid, carbonic acid is formed. The acid cannot be isolated from its solution and is probably $CO_2.H_2O$.

$$H_2O(l) + CO_2(g) \rightleftharpoons H^+(aq) + HCO_3^-(aq)$$

3   Alkalis forms carbonates and hydrogencarbonates.

$$2OH^-(aq) + CO_2(g) \longrightarrow CO_3^{2-}(aq) + H_2O(l)$$
$$\text{carbonate}$$

$$OH^-(aq) + CO_2(g) \longrightarrow HCO_3^-(aq)$$
$$\text{excess} \qquad\qquad \text{hydrogencarbonate}$$

Salts are hydrolysed in water making their solutions weakly acidic.

$$CO_3^{2-}(aq) + H_2O(l) \rightleftharpoons OH^-(aq) + HCO_3^-(aq)$$
$$HCO_3^-(aq) + H_2O(l) \rightleftharpoons OH^-(aq) + H_2CO_3(aq)$$

4   **Metallic carbonates** are precipitated when sodium carbonate solution is added to a solution of a soluble metal salt, except Na_2CO_3 and K_2CO_3 which are soluble. The metals magnesium, lead and zinc form basic carbonates with sodium carbonates and so sodium hydrogencarbonate is needed, e.g.

$$Mg^{2+}(aq) + 2HCO_3^-(aq) \longrightarrow MgCO_3(s) + H_2O(l) + CO_2(g)$$

Copper does not form a normal carbonate and weak bases precipitate the corresponding hydroxide and carbon dioxide, e.g. the metals aluminium and iron(III) form hydroxides.

$$2Al^{3+}(aq) + 3CO_3^{2-}(aq) + 3H_2O(l) \longrightarrow 2Al(OH)_3(s) + 3CO_2(g)$$

On heating, carbonates form oxides and carbon dioxide – Na_2CO_3 and K_2CO_3 are not decomposed by heat.

5   **Hydrogencarbonates**   Only two common hydrogencarbonates are solids, $KHCO_3$ and $NaHCO_3$. Others are known only in solution and decompose to carbonates if the water is allowed to evaporate. On heating carbonates or oxides are formed. Calcium hydrogencarbonate is important geologically.

6   **To identify carbonates and hydrogencarbonates**   On the addition of magnesium sulphate solution, a white precipitate formed

immediately indicates a normal carbonate, a white precipitate on boiling indicates a hydrogencarbonate.

7   To identify carbon dioxide   Pass the gas through clear lime water, the formation of a white precipitate (cloudiness) indicates carbon dioxide. The lime water becomes clear if excess carbon dioxide is passed because soluble calcium hydrogencarbonate is formed.

$$CO_2(g) + Ca(OH)_2(aq) \rightleftharpoons CaCO_3(s) + H_2O(l)$$

$$CaCO_3(s) + H_2O(l) + CO_2(g) \rightleftharpoons Ca(HCO_3)_2(aq)$$

8   Photosynthesis   The formation of sugars in plants by the combination of carbon dioxide and water in the presence of chlorophyll and sunlight. A very simplified representation is:

$$6H_2O(l) + 6CO_2(g) \xrightarrow[\text{sunlight}]{\text{chlorophyll}} C_6H_{12}O_6(s) + 6O_2(g)$$

Carbon disulphide, CS_2

$S{=}C{=}S$, is obtained by direct combination when sulphur vapour is passed over **hot charcoal**. Commercially the coke is heated in an electric furnace by producer gas and an arc is maintained between two carbon electrodes.

$$C(s) + 2S(s) \longrightarrow CS_2(g)$$

It is purified by fractional distillation, then shaken with mercury and distilled over phosphorus(V) oxide.

Properties
1   Colourless, mobile liquid with a sweet smell when pure but usually an evil smell when slightly impure.
2   Mixes slightly with water and readily with alcohol and benzene. It is a good solvent for sulphur, phosphorus and iodine.
3   Dangerously inflammable. Ignites in air or oxygen burning with a blue flame.

$$CS_2(l) + 3O_2(g) \longrightarrow CO_2(g) + 2SO_2(g)$$

4   A mixture of nitrogen oxide and carbon disulphide burns with a brilliant flash.

$$CS_2(g) + 10NO(g) \longrightarrow 2CO(g) + 4SO_2(g) + 5N_2(g)$$

5   It is the sulphur analogue of carbon dioxide and forms trithiocarbonates derived from trithiocarbonic acid, H_2CS_3. If shaken with sodium sulphide solution, it forms sodium trithiocarbonate.

$$Na_2S(aq) + CS_2(l) \longrightarrow Na_2CS_3(aq)$$

6   Forms tetrachloromethane with chlorine.

Uses
1   As a solvent for fats, oils, rubber, sulphur, bromine and is used as a solvent for extraction.
2   Used in the production of viscose silk.
3   Used in the production of glossy cellulose.

Tetrachloromethane (carbon tetrachloride), CCl_4, is obtained by passing chlorine through carbon disulphide boiled under reflux in the presence of iodine as catalyst.

$$CS_2(l) + 3Cl_2(g) \longrightarrow CCl_4(l) + S_2Cl_2(l)$$

Tetrachloromethane and disulphur dichloride are separated by fractional distillation, the carbon disulphide boiling off first.

Properties
1   Colourless liquid, good solvent for fats and non-flammable.
2   Rather inert chemically (*c.f.* $SiCl_4$).

Uses
1   As a solvent, cleaning and de-greasing agent.
2   Used as pyrene in fire extinguishers but is oxidized to the very poisonous $COCl_2$ when it comes into contact with very hot surfaces.
3   Reduced by boiling with iron/water to trichloromethane (chloroform).

$$CCl_4(l) + Fe(s) + H^+(aq) \longrightarrow Fe^{2+}(aq) + Cl^- + CHCl_3(l)$$

Cyanogen, $(CN)_2$ **Very poisonous gas** made by mixing a warm concentrated solution of potassium cyanide and copper(II) sulphate.

$$2Cu^{2+}(aq) + 4CN^-(aq) \longrightarrow 2CuCN(s) + (CN)_2(g)$$

Cyanogen shows some similarities with iodine and is called a pseudo-halogen.

Hydrogen cyanide (hydrocyanic acid, prussic acid), HCN. This is a **very dangerous poison** and should not be handled by students. It is formed by distilling potassium cyanide and a mixture of concentrated sulphuric acid and an equal volume of water.

$$KCN(s) + H_2SO_4(l) \longrightarrow KHSO_4(s) + HCN(l)$$

Properties
1   Colourless, very poisonous liquid smelling of bitter almonds.
2   Burns with a purple flame.

$$4HCN(g) + 5O_2(g) \longrightarrow 2H_2O(l) + CO_2(g) + 2N_2(g)$$

3   Very weak acid and feebly ionized in water.

$$HCN(aq) \rightleftharpoons H^+(aq) + CN^-(aq)$$

Forms cyanides with caustic alkalis.

$$HCN(aq) + KOH(aq) \longrightarrow KCN(aq) + H_2O(l)$$

Cyanides are so weak that they are decomposed by carbon dioxide.

$$2CN^-(aq) + CO_2(aq) + H_2O(l) \longrightarrow CO_3^{2-}(aq) + 2HCN(aq)$$

They are considerably hydrolysed and their solutions smell of hydrogen cyanide. Many cyanides of the heavier metals are insoluble in water. Complex metallo cyanide ions occur frequently, e.g. the cyanides of silver, copper(I) and cadmium dissolve in excess potassium cyanide.

$$Ag(CN)(s) + CN^-(aq) \rightleftharpoons Ag(CN)_2^-(aq)$$

$$Cu(CN)(s) + CN^-(aq) \rightleftharpoons Cu(CN)_2^-(aq)$$

$$Cd(CN)_2 + 2CN^-(aq) \rightleftharpoons Cd(CN)_4^{2-}(aq)$$

4 Hydrolysed in cold water to form ammonium methanoate.

$$HCN(aq) + 2H_2O(l) \longrightarrow HCOO^-(aq) + NH_4^+(aq)$$

5 Reduced by Zn/HCl to methylamine.

$$2Zn(s) + 4H^+(aq) + HCN(aq) \longrightarrow 2Zn^{2+}(aq) + CH_3NH_2(aq)$$

6 Forms addition compounds with aldehydes and ketones – cyanohydrins.

$$CH_3-C\overset{\displaystyle H}{\underset{\displaystyle O}{}} + HCN \longrightarrow CH_3-\underset{\displaystyle CN}{\overset{\displaystyle H}{C}}-OH$$

ethanal   2 hydroxypropanonitrile

Uses

1 Fumigating ships and buildings.

2 Potassium and sodium cyanides used in the extraction of gold.

Cyanates Obtained when cyanides are fused with metallic oxides, cyanates are very stable.

$$Pb_3O_4(s) + 4KCN(s) \longrightarrow 3Pb(s) + 4KCNO(s)$$

Cyanic acid, HCNO, is a colourless gas which liquefies in a freezing mixture. On standing, it polymerizes to cyanuric acid, $H_3C_3N_3O_3$ and the aqueous solution decomposes rapidly.

$$HCNO(aq) + 2H_2O(l) \longrightarrow NH_4^+(aq) + HCO_3^-(aq)$$

On heating, ammonium cyanate forms urea.

Carbon cycle

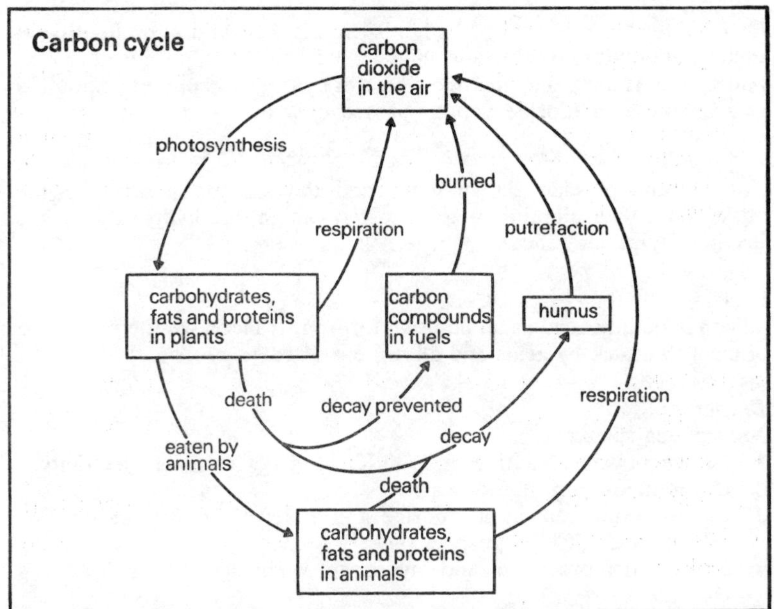

Silicon

Occurrence   Silicon is the second most abundant element (after oxygen) in the Earth's crust. Found as silica, SiO_2 as quartz, sand and rock crystal, as metallic silicates with complex structures such as felspar $K_2Al_2Si_6O_{16}$ and kaolinite $Al_2Si_2O_7.2H_2O$.

Extraction   **Amorphous silicon** is obtained by heating coke and excess sand in an electric furnace.

$$SiO_2(s) + 2C(s) \longrightarrow Si(s) + 2CO(g)$$

Excess sand is needed to avoid the formation of carborundum.

$$SiO_2(s) + 3C(s) \longrightarrow SiC(s) + 2CO(g)$$

Preparation   Amorphous silicon is obtained when silica and powdered magnesium are very strongly heated in the absence of air.

$$2Mg(s) + SiO_2(s) \longrightarrow Si(s) + 2MgO(s)$$

$$2Mg(s) + SiO_2(s) \longrightarrow Mg_2Si(s) + 2MgO(s)$$

The oxide and silicide are removed by hot hydrochloric acid leaving the silicon as a brown powder.

Crystalline silicon is obtained when silicon is dissolved in molten metal such as aluminium or zinc and then cooled. It may be produced *in situ*. Aluminium is melted in an atmosphere of hydrogen to prevent oxidation and silicon tetrachloride vapour passed over it.

$$3SiCl_4(g) + 4Al(l) \longrightarrow 3Si(s) + 4AlCl_3(g)$$

The aluminium chloride sublimes and the silicon dissolves in the aluminium. The aluminium is dissolved out in hot hydrochloric acid leaving crystalline silicon.

Uses

Silicon is mainly used as an additive for iron. It increases the resistance of steel to attack by acids and alloys, e.g. hardens bronze.

Properties

Amorphous silicon

1 Brown powder, melting at 1700 K, does not conduct electricity.
2 Burns in oxygen at red-heat.
3 Ignites spontaneously in fluorine and at 730 K in chlorine forming $SiF_4(g)$ and $SiCl_4(g)$ respectively.
4 Forms trichlorosilane and hydrogen when heated in hydrogen chloride.

$$Si(s) + 3HCl(g) \longrightarrow SiHCl_3(l) + H_2(g)$$

5 Silicon is attacked slowly by steam.

$$Si(s) + 2H_2O(g) \longrightarrow SiO_2(s) + 2H_2(g)$$

6 Resistant to attack by acids and only hydrogen fluoride has any appreciable effect.

$$Si(s) + 6HF(l) \longrightarrow H_2SiF_6(l) + 2H_2(g)$$

7 Dissolves readily in hot concentrated caustic alkali.

$$Si(s) + 2OH^-(aq) + H_2O(l) \longrightarrow SiO_3^{2-}(aq) + 2H_2(g)$$

8 When heated in nitrogen, silicon forms silicon nitride.

$$3Si(s) + 2N_2(g) \longrightarrow Si_3N_4(s)$$

9 Silicon reacts with the more reactive metals to form silicides.

$$Si(s) + 2Mg(l) \longrightarrow Mg_2Si(s)$$

Crystalline silicon

Dark-grey, very hard solid of variable density. It conducts electricity but is less reactive than amorphous silicon. It is used in semi-conductors.

Compounds of silicon

Hydrides, silanes S_nH_{2n+2}, e.g. silane SiH_4, disilane (silicoethane) Si_2H_6, trisilane (silico-propane) Si_3H_8. Ten known.

Preparation A mixture of silanes is obtained by the action of 20% hydrochloric acid on magnesium silicide. Monosilane can be obtained by fractionation *in vacuo* (the gas is spontaneously inflammable).

$$Mg_2Si(s) + 4HCl(aq) \longrightarrow 2MgCl_2(aq) + SiH_4(g)$$

It is also formed by the action of tetrahydridoaluminate in ether on silicon tetrahalide.

Properties
1 Silane is a colourless gas, spontaneously inflammable in air.

$$SiH_4(g) + 2O_2(g) \longrightarrow SiO_2(s) + 2H_2O(g)$$

2 Reacts with caustic alkalis (*c.f.* methane – no reaction).

$$SiH_4(g) + 2OH^-(aq) + H_2O(l) \longrightarrow SiO_3^{2-}(aq) + 4H_2(g)$$

3 Strong reducing agent, changing iron(III) to iron(II) and precipitates silver from solutions of silver salts.

$$4Fe^{3+}(aq) + SiH_4(g) \longrightarrow Si(s) + 4H^+(aq) + 4Fe^{2+}(aq)$$

$$4Ag^+(aq) + SiH_4(g) \longrightarrow Si(s) + 4H^+(aq) + 4Ag(s)$$

(*c.f.* methane which reduces some metallic oxides, e.g.

$$4CuO(s) + CH_4(g) \longrightarrow 4Cu(s) + CO_2(g) + 2H_2O(g)$$

4 Silane is less stable than methane and is decomposed at 670 K.
5 Undergoes substitution by chlorine and reactions are more vigorous than for methane – SiH_3Cl; SiH_2Cl_2; $SiHCl_3$; $SiCl_4$.

$$SiH_4(g) + Cl_2(g) \longrightarrow SiH_3Cl(g) + HCl(g)$$

6 No multiple bond silanes like ethyne.
7 Silyl group is similar to methyl group but, for example, trisilylamine $(SiH_3)_3N$, the analogue of trimethylamine $(CH_3)_3N$ is not basic and is not a donor molecule.

Halides

Silicon tetrafluoride, SiF_4, is obtained by heating hydrogen fluoride (actually a mixture of calcium fluoride and concentrated sulphuric acid) and silica.

$$CaF_2(s) + H_2SO_4(l) \longrightarrow CaSO_4(s) + 2HF(l)$$

$$SiO_2(s) + 4HF(l) \longrightarrow SiF_4(g) + 2H_2O(l)$$

Silane is a colourless gas with a pungent smell. It can be solidified by cooling at atmospheric pressure without it liquefying first. Forms silicon when heated with sodium or potassium.

$$3SiF_4(g) + 4K(l) \longrightarrow Si(s) + 4KF(s)$$

Hexafluorosilicic acid, H_2SiF_6, is formed with colourless, gelatinous silicic acid, H_2SiO_2, when **silicon tetrafluoride** is passed into **water**.

$$3SiF_4(g) + 3H_2O(l) \longrightarrow H_2SiO_2(s) + 2H_2SiF_6(aq)$$

Hexafluorosilicic acid is obtained only as a dihydrate $H_2SiF_6.2H_2O$. It is a weak acid and its salts are used in electroplating and in producing lead of a very high purity.

Silicon tetrachloride, $SiCl_4$, is formed by **heating silicon in chlorine** or by heating a mixture of silica and carbon in a stream of chlorine.

$$SiO_2(s) + 2C(s) + 2Cl_2(g) \longrightarrow SiCl_4(g) + 2CO(g)$$

Colourless liquid, fuming in moist air and hydrolysed by water (*c.f.* tetrachloromethane).

$$SiCl_4(l) + 3H_2O(l) \longrightarrow H_2SiO_3(s) + 4HCl(g)$$

It is decomposed when heated with alkali metals.

$$SiCl_4(l) + 4K(s) \longrightarrow Si(s) + 4KCl(s)$$

There is no $SiCl_6^{2-}$ ion corresponding to the octahedral SiF_6^{2-}.

Oxides
Silicon(II) oxide, SiO, is obtained by the **reduction of silica** at high temperature but it is unstable at room temperature.

$$SiO_2(s) + Si(s) \longrightarrow 2SiO$$

Silica(IV) oxide, silica, SiO_2. This has a giant covalent structure (*c.f.* the simple molecular structure of carbon dioxide). Each silicon atom is surrounded by four oxygen atoms at the apices of a tetrahedron. When quartz is melted and then allowed to solidify, it forms a glass with a low coefficient of expansion. Quartz glass is transparent to U.V. light and is used in mercury lamps. **Amorphous** silica occurs as flint.

Uses
1 Sand is used in mortar, cement and concrete.
2 Quartz is used in electrical instruments.
3 Manufacture of glass.

Properties Amorphous silica is more reactive than the crystalline forms.

1   Slightly soluble if heated with water under pressure.

$$H_2O(l) + SiO_2(s) \longrightarrow H_2SiO_3(aq)$$

2   Weakly acidic and reacts rapidly with hot concentrated or fused alkali to form silicates.

$$SiO_2(s) + 2OH^-(aq) \longrightarrow SiO_3^{2-}(aq) + H_2O(l)$$

3   It displaces other acidic anhydrides to form silicates because silica is non-volatile when other oxides have vapourized.

$$Na_2SO_4(s) + SiO_2(s) \longrightarrow Na_2SiO_3(l) + SO_3(g)$$

$$Na_2CO_3(s) + SiO_2(s) \longrightarrow Na_2SiO_3(l) + CO_2(g)$$

4   Forms silicon carbide or carborundum when heated with carbon in an electric furnace.

$$SiO_2(s) + 3C(s) \longrightarrow SiC(s) + 2CO(g)$$

5   Silica is reduced when heated with powdered magnesium to amorphous silicon.

$$SiO_2(s) + 2Mg(s) \longrightarrow Si(s) + 2MgO(s)$$

If excess magnesium is used, magnesium silicide is formed, Mg_2Si.

Silicic acids
There are two simple acids from which most silicates derive, tetraoxosilicic acid (orthosilicic acid) $Si(OH)_4$ or H_2SiO_4 and trioxosilicic acid (metasilicic acid) $O{=}Si(OH)_2$ or H_2SiO_3.

Structure of silicates
1   Three dimensional lattice as in quartz.
2   Extended anions which may form chains $(SiO_3)_n^{2n}$, double chains $(Si_4O_{11})_n^{6n-}$ or sheets $(Si_2O_5)^{2n-}$.
3   Discrete anions; individual tetrahedral anions in which one or two oxygen atoms are shared by each tetrahedron, e.g. SiO_4^{4-} orthosilicates, $Si_2O_7^{6-}$ pyrosilicates and $Si_3O_9^{6-}$ cyclic silicates.

Silicones
Hydrolysis of alkylchlorosilanes yields molecules which can undergo condensation polymerization.

$$HOSi(CH_3)_2OH + HO(CH_3)_2SiOH \longrightarrow$$
$$H_2O + HOSi(CH_3)_2{-}O{-}(CH_3)_2SiOH$$

The length of the chains and the degree of cross-linking determines the properties of the silicones which may be oils or rubber-like solids. Used in paints, varnishes and in proofing textiles.

Similarities between silicon and boron
1 Oxides B_2O_3 and SiO_2 are both acidic.
2 Both form oxyacids, H_3BO_3 and H_2SiO_3 – borates and silicates.
3 Both form volatile fluorides, BF_3 and SiF_4, which are hydrolysed to HBO_2 and HBF_4, H_2SiO_3 and H_2SiF_4 respectively.
4 Form metallic borides and silicides which yield hydrides with mineral acids.
5 Both form a series of hydrides.
6 Both form glassy materials.

Germanium, Ge

A grey brittle solid with semi-conducting properties so that it is used in transistors. It forms compounds containing the $+2$ and $+4$ oxidation states. It reacts readily with dilute alkalis.

$$Ge(s) + 2OH^-(aq) + H_2O(l) \longrightarrow GeO_3^{2-}(aq) + 2H_2(g)$$

Hydrides Forms several unstable germanes, Ge_nH_{2n+2}.

Oxides **Germanium(II) oxide**, GeO, is more stable than silicon(II) oxide – germanium is more metallic than silicon.

$$GeCl_2(s) + H_2O(l) \longrightarrow GeO(s) + 2HCl(aq)$$

Anhydrous germanium oxide is a black powder. When hydrated it is yellow. It is amphoteric but more acidic than tin(II) oxide and lead(II) oxide. **Germanium(IV) oxide, germanium dioxide**, GeO_2, is formed by the hydrolysis of germanium(IV) chloride $GeCl_4$. It is acidic (both tin(IV) oxide and lead(IV) oxide are amphoteric).

Chlorides **Germanium(II) chloride**, $GeCl_2$, is formed when germanium(IV) chloride vapour is passed over germanium. It is readily hydrolysed by water. **Germanium(IV) chloride**, $GeCl_4$, is obtained by passing chlorine over germanium. It is hydrolysed by water – more slowly than silicon(IV) chloride but more rapidly than tin(IV) chloride and lead(IV) chloride.

Tin, Sn

Occurrence Main ore is cassiterite or tinstone, SnO_2, found in lodes or veins when it is mined, or in alluvial deposits when it is dredged.

Extraction Ore is **concentrated** by crushing and washing to remove less dense earthy material. The residue is roasted to remove volatile oxides of, e.g. arsenic and sulphur. Iron compounds are removed by electro-magnetic separation. The concentrated tin ore is **smelted**. It is heated with **carbon** in a reverberatory furnace at 1500 K. The oxide is reduced to the metal which is molten and runs to the bottom of the furnace and is tapped off.

$$SnO_2(s) + 2C(s) \longrightarrow Sn(l) + 2CO(g)$$

The tin is **purified** by re-melting the tin, which is run off leaving less fusible material. The molten metal is then stirred in contact with air to oxidize impurities which can be skimmed from the surface.

Physical properties

1 White metal, soft and ductile – made into wires and tinfoil.
2 When a bar of tin is bent it 'cries' the crystals grating against each other.
3 Three allotropes Grey $\rightleftharpoons$ white $\rightleftharpoons$ rhombic.
   Grey tin is stable below 286 K and has a gigantic diamond-like lattice. **White tin** is stable up to 434 K and consists of tetragonal crystals. **Rhombic tin** is stable from 434 K up to the melting point of the tin.

Chemical properties

1 Not affected by oxygen at room temperature but it burns to tin(IV) oxide.
2 Dilute acids attack tin only slowly if at all. Hot dilute, or concentrated **hydrochloric acid** slowly forms tin(II) chloride.

$$Sn(s) + 2H^+(aq) \longrightarrow Sn^{2+}(aq) + H_2(g)$$

Dilute **sulphuric acid** has little effect, the hot concentrated acid forms mainly tin(IV) sulphate and also basic sulphate and sulphur. The main reaction is

$$Sn(s) + 4H_2SO_4(l) \longrightarrow Sn(SO_4)_2(aq) + 4H_2O(l) + 2SO_2(g)$$

Nitric acid has little action forming tin(II) nitrate, oxides of nitrogen and ammonium nitrate. The concentrated acid attacks tin readily forming nitrogen dioxide and a white precipitate of hydrated oxide (metastannic acid).

$$Sn(s) + 4HNO_3(conc.aq.) \longrightarrow SnO_2(s) + 4NO_2 + 2H_2O(l)$$

3 Hot concentrated caustic alkalis give hydrogen and hexadroxostannate(IV) in solution. The reaction is slow.

$$Sn(s) + 2OH^-(aq) + 4H_2O(l) \longrightarrow [Sn(OH)_6]^{2-}(aq) + 2H_2(g)$$

4 Finely divided tin dissolves in liquid ammonia forming red NH_4Sn_9 containing the polyanion $(Sn_9)^{4-}$. Lead behaves similarly.
5 **Tin(II) chloride** is formed by the action of dry **hydrogen chloride** on heated tin. It is a white ionic solid.

$$Sn(s) + 2HCl(g) \longrightarrow SnCl_2(s) + H_2(g)$$

6   Tin(IV) chloride is formed by passing dry chlorine over heated tin. It is a colourless covalent liquid.

$$Sn(s) + 2Cl_2(g) \longrightarrow SnCl_4(l)$$

Uses
1   As a coating on steel, tin-plate, it resists corrosion and is widely used in canning. If scratched, because tin is less electropositive than iron, the oxidation of iron is accelerated (*c.f.* Zn/Fe in galvanishing).
2   In alloys, e.g. solder (Sn/Pb), pewter (Sn/Pb/Sb), coinage bronze (Cu/Sn/Zn).

Compounds of tin

Tin shows the oxidation states +2 and +4. Compounds containing tin(II) are generally ionic with tin behaving as a metal. Compounds containing tin(IV) are generally covalent.

Tin(IV) compounds (stannate(IV))
Stannane, SnH_4.   As a heavy metal, tin would not be expected to form an ionic hydride like that of NaH. The covalent hydride SnH_4 exists but is unstable. It is formed by the reduction of tin(IV) chloride by lithium tetrahydridoaluminate in ether.

$$SnCl_4(l) + LiAlH_4(s) \longrightarrow SnH_4(g) + LiAlCl_4(s)$$

It decomposes to deposit a tin mirror.

Tin(IV) oxide, SnO_2, is formed as hydrated tin(II) oxide, a white gelatinous precipitate of uncertain composition, when ammonia solution is added to tin(IV) chloride solution.

$$Sn^{4+}(aq) + 4OH^-(aq) \longrightarrow (Sn)_2.2H_2O(s)$$

Aqueous alkali also forms a white precipitate with tin(IV) chloride but the precipitate dissolves in excess alkali forming hexahydroxostannate(IV).

$$SnO_2.2H_2O(s) + 2OH^-(aq) \longrightarrow [Sn(OH)_6]^{2-}(aq)$$

This oxide has no basic properties and tin is showing non-metallic properties. If dilute acid is added to the solution of $Na_2[Sn(OH)_6]$ a white precipitate is formed which was thought to be stannic acid but it is probable that the various precipitates are actually tin(II) oxide with different degrees of hydration and particle size.

Tin(IV) chloride, $SnCl_4$. Formed by the action of dry chlorine on heated tin.

$$Sn(s) + 2Cl_2(g) \longrightarrow SnCl_4(l)$$

It is a volatile, covalent liquid which fumes in moist air. Tin is behaving as a non-metal (*c.f.* CCl_4 and $SiCl_4$). It is hydrolysed by water.

$$SnCl_4 + 4H_2O(l) \longrightarrow SnO_2.2H_2O(s) + 4HCl(aq)$$

It is soluble in organic solvents (characteristic of covalent compounds) but is soluble in water and is hydrated (characteristic of ionic compounds). Therefore the hydrated Sn^{4+} must be formed in water and undergo hydrolysis (*c.f.* aluminium).

$$[Sn.xH_2O]^{4+} \rightleftharpoons [SnOH.(x-1)H_2O]^{3+} + H^+ \rightleftharpoons$$
$$[Sn(OH)_2.(x-2)H_2O]^{2+} + 2H^+$$

If alkali is present, the final product is $[Sn(OH)_6]^{2-}$. The hydrolysis can be suppressed by hydrochloric acid.

Tin(IV) sulphide, SnS_2. A yellow solid precipitated when hydrogen sulphide is passed into a hot solution of tin(IV) chloride acidified with dilute hydrochloric acid.

$$SnCl_4(aq) + 2H_2S(g) \longrightarrow SnS_2(s) + 4HCl(aq)$$

It is soluble in alkali forming hydroxostannate(IV) and trithiostannate(IV).

$$3SnS_2 + 6OH^-(aq) \longrightarrow [Sn(OH)_6]^{2-}(aq) + 2SnS_3^{2-}(aq)$$

It is soluble in solutions containing S^{2-}.

$$SnS_2 + 2S^{2-} \longrightarrow SnS_3^{2-}$$
$$SnS_2 + 2s^{2-} \longrightarrow SnS_4^{4-}$$

Tin(II) compounds (stannite II)
Tin(II) oxide, SnO. The addition of a small amount of **alkali** to the solution of tin(II) salt precipitates the **hydroxide** which when filtered off and heated forms the oxide. It is obtained by heating **tin(II) ethanedioate**.

$$SnC_2O_4(s) \longrightarrow SnO(s) + CO(g) + CO_2(g)$$

Dark-grey or brown powder, rapidly oxidized to tin(IV) oxide. It is amphoteric.

Tin(II) hydroxide, $Sn(OH)_2$. Obtained as a gelatinous white precipitate of indefinite composition by the addition of **alkali** to a solution of a **tin(II) salt**. It dissolves in excess hydroxide.

$$Sn^{2+}(aq) + 2OH^-(aq) \longrightarrow Sn(OH)_2(s)$$

It is amphoteric.

Tin(II) chloride, $SnCl_2$. **Tin** reacts with moderately concentrated **hydrochloric acid** to form the dihydrate $SnCl_2.2H_2O$.

It is hydrolysed readily, precipitating a basic chloride or even tin(II) hydroxide in excess water. It is obtained by heating tin in a current of dry hydrogen chloride.

$$SnCl_2(s) + H_2O(l) \rightleftharpoons Sn(OH)Cl(s) + HCl(aq)$$

$$SnCl_2(s) + 2H_2O(l) \rightleftharpoons Sn(OH)_2(s) + 2HCl(aq)$$

Tin(II) chloride is usually stored with a little concentrated hydrochloric acid and a few pieces of tin. Tin(II) chloride has strong reducing properties in which Sn^{2+} acts as an electron donor.

$$Sn^{2+}(aq) \longrightarrow Sn^{4+}(aq) + 2e^-$$

The tin(IV) ion probably exists only in a complex ion which is hydrated. Tin(II) chloride reduces:

1 **Mercury(II) to mercury(I)** and finally **grey metallic** mercury may be precipitated.

$$ZHg^{2+}(aq) + Sn^{2+}(aq) + 2Cl^-(aq) \longrightarrow Hg_2Cl_2(s) + Sn^{4+}(aq)$$
$$Hg^{2+}(s) + Sn^{2+}(aq) \longrightarrow 2Hg(l) + Sn^{4+}(aq)$$

2 **Iron(III) to iron(II)**.

$$2Fe^{3+}(aq) + Sn^{2+}(aq) \longrightarrow 2Fe^{2+}(aq) + Sn^{4+}(aq)$$

3 **Acidified potassium manganate(VII)** (purple) to **potassium manganate(II)** (colourless).

$$2MnO_4^-(aq) + 5Sn^{2+}(aq) + 16H^+(aq) \longrightarrow$$
$$2Mn^{2+}(aq) + 5Sn^{4+}(aq) + 8H_2O(l)$$

4 **Acidified potassium dichromate(VI)** (orange) to **potassium chromate(III)**.

$$Cl_2O_7{}^{2-}(aq) + 3Sn^{2+}(aq) + 14H^+(aq) \longrightarrow$$
$$2Cr^{3+}(aq) + 3Sn^{4+}(aq) + 7H_2O(l)$$

5 **Iodine** dissolved in potassium iodine solution to **iodide**.

$$Sn^{2+}(aq) + I_2 \longrightarrow Sn^{4+}(aq) + 2I^-(aq)$$

6 **Nitrobenzene** in the presence of hydrochloric acid to **aniline hydrochloride**

$$C_6H_5NO_2 + 3Sn^{2+}(aq) + 6H^+(aq) \longrightarrow$$
$$C_6H_5NH_2 + 3Sn^{4+} + 2H_2O(l)$$

Tin chloride is oxidized in air but this is prevented by storing with a few pieces of tin.

Tin(II) sulphide, SnS_2. Formed as a brown precipitate when hydrogen sulphide is passed through the solution of a tin(II) salt. It is not soluble in the presence of S^{2-} but if sulphur is present, then free sulphur adds on.

$$SnS(s) + S(s) \longrightarrow SnS_2$$

The tin(IV) sulphide then dissolves.

Lead, Pb

Occurrence Galena, PbS, lead(II) sulphide and also as the **carbonate** and **sulphate**.

Extraction The washed ore is **concentrated** by flotation. It is agitated with water containing chemicals which cause frothing and the galena particles are held in the froth. The concentrated ore is then **roasted** in a reverberatory furnace in a plentiful supply of air when lead(II) oxide is formed and sulphur dioxide comes off.

$$2PbS(s) + 3O_2(g) \longrightarrow 2PbO(s) + 2SO_2(g)$$

The oxide is **smelted**. It is heated with **coke** in a small blast furnace when it is reduced to lead.

$$PbO(s) + C(s) \longrightarrow Pb(l) + CO(g)$$

Iron is introduced into the furnace to act as a reducing agent on any remaining lead(II) sulphide.

$$PbS(s) + Fe(s) \longrightarrow Pb(l) + FeS(s)$$

Limestone is also added to act as a flux, changing sand materials to fusible calcium silicate.

$$CaCO_3(s) + SiO_2(s) \longrightarrow CaSiO_3(l) + CO_2(g)$$

The lead is **refined** by heating it in a shallow reverberatory furnace or by blowing air through the molten metal. Impurities are oxidized and float to the surface from which they can be skimmed.

Pure lead is obtained by the electrolysis of hexafluorosilicic acid, H_2SiF_6 and the lead salt Pb_2SiF_6 with a little gelatin. The cathode is pure lead and the impure lead is made the anode of the cell.

Cathode $Pb^{2+} + 2e^- \longrightarrow Pb(s)$

Anode $Pb - 2e^- \longrightarrow Pb^{2+}(aq)$ Any silver or gold present precipitates in the anode slime.

The gelatin ensures that a smooth coherent deposit of lead is laid down.

Uses

1 In domestic water pipes (it is attacked by soft water but not by hard water) in guttering and roofing.

2 As a lining for chemical reaction vessels, it is resistant to attack, including attack by sulphuric acid.
3 In accumulators.
4 In alloys: pewter and solder (lead and tin); lead shot (lead and arsenic); pewter (lead, tin and antimony).
5 Sheets, as protection against radiation and also as lead glass.

Physical properties Bluish grey, lustrous, soft metal. Soft enough to be cut with a knife, can be rolled into sheets but it is not strong enough for use as wire.

Chemical properties
1 Becomes covered with a dull, thin layer of hydroxide and carbonate. When heated to just above its melting point it forms yellow lead(II) oxide (massicot). At a higher temperature it forms red lead, Pb_3O_4.

$$2Pb(l) + O_2(g) \longrightarrow 2PbO(s)$$

$$6PbO(s) + O_2(g) \longrightarrow 2Pb_3O_4(s)$$

Finely divided lead is pyrophoric.

2 Pure water, free from air, does not attack lead. In the presence of air, sparingly soluble hydroxide is formed.

$$2Pb(s) + 2H_2O(l) + O_2(g) \longrightarrow 2Pb(OH)_2(s)$$

In the presence of carbonate or sulphate ions, as in hard water, a protective layer of lead carbonate or sulphate is formed. Lead is a dangerous insidious poison and lead pipes cannot be used for soft waters.

3 Lead is attacked by hot concentrated **hydrochloric acid** forming insoluble lead(II) chloride.

$$Pb(s) + 2H^+(aq) \longrightarrow Pb^{2+}(aq) + H_2(g)$$

4 **Sulphuric acid** attacks lead only if the acid is hot and concentrated.

$$Pb(s) + 2H_2SO_4(l) \longrightarrow PbSO_4(s) + 2H_2O(l) + SO_2(g)$$

5 **Nitric acid** attacks lead to form lead(II) nitrate and oxides of nitrogen. Pure nitric acid has little effect.

6 Hot concentrated solutions of **alkalis** very slowly form hexahydroxoplumbates(II).

Compounds of lead

In the +2 oxidation state, lead acts as a typical metal forming ionic compounds. In the +4 oxidation state lead behaves as a non-metal forming covalent compounds resembling carbon and silicon.

Lead(IV) compounds

Hydride   The stability of hydrides decreases down group IV and only one is known for lead, **plumbane**, PbH_4. It is unstable, decomposing rapidly.

Lead tetraethyl, $Pb(C_2H_5)_4$, is a covalent compound, but it is very stable and is used as an anti-knock in petrol. It is obtained by heating lead, sodium and chloroethane.

$$Pb + 4Na + 4C_2H_5Cl \longrightarrow Pb(C_2H_5)_4 + 4NaCl$$

Lead(IV) oxide, PbO_2.   Covalent $O{=}Pb{=}O$ and not a salt of hydrogen peroxide (*c.f.* barium peroxide, $Ba^{2+}O_2^{2-}$). It does not form hydrogen peroxide with dilute acids.

Preparation   Action of excess dilute nitric acid on red lead forms a dark-brown precipitate of lead(II) oxide.

$$Pb_3O_4(s) + 4HNO_3(aq) \longrightarrow PbO_2(s) + 2Pb(NO_3)_2(aq) + 2H_2O(l)$$

Properties

1   Decomposes on heating above 373 K to lead(II) oxide.

$$2PbO_2(s) \longrightarrow 2PbO(s) + O_2(g)$$

2   Shows amphoteric character. At 273 K, it will form lead(IV) chloride with concentrated hydrochloric acid.

$$PbO_2(s) + 4HCl(conc) \longrightarrow PbCl_4(l) + 2H_2O(l)$$

Lead(IV) oxide is behaving as a basic oxide. At higher temperatures it reacts with concentrated hydrochloric acid to form chlorine and lead(II) chloride.

$$PbO_2(s) + 4HCl(conc) \longrightarrow 2H_2O(l) + PbCl_2(s) + Cl_2(g)$$

3   Reacts slowly with alkali fused in solution to form hydroxoplumbate(IV). Here it is behaving as an acidic oxide.

$$PbO_2(s) + 2OH^-(aq) + 2H_2O(l) \longrightarrow [Pb(OH)_6]^{2-}(aq)$$

On heating this forms plumbate, e.g.

$$Na_2Pb(OH)_6 \longrightarrow Na_2PbO_3 + 3H_2O$$

4   **Oxidizing properties**
Oxidizes concentrated hydrochloric acid to chlorine (see above). Changes sulphur dioxide to lead(II) sulphate. Used as an oxidizing substance in the head of matches. Ignites a stream of hydrogen sulphide. (*c.f.* SnO_2 – difference in effect of inert pair).

Lead(IV) chloride, $PbCl_4$

Preparation Lead(IV) oxide slowly added to ice-cold concentrated hydrochloric acid. When filtered, the yellow liquid contains the complex hexachloroplumbate(IV) ion.

$$PbO_2(s) + 4HCl(conc) \longrightarrow PbCl_4(l) + 2H_2O(l)$$

$$PbCl_4(l) + 2Cl^-(aq) \longrightarrow PbCl_6^{2-}(aq)$$

Addition of ammonium chloride gives the complex salt ammonium hexachloroplumbate(IV) as a yellow precipitate.

$$2NH_4Cl(aq) + PbCl_4(l) \longrightarrow (NH_4)_2PbCl_6(s)$$

When this precipitate is treated with cold concentrated sulphuric acid, oily lead(IV) oxide separates.

$$(NH_4)_2PbCl_6(s) + H_2SO_4(conc) \longrightarrow$$
$$(NH_4)_2SO_4(s) + PbCl_4(l) + 2HCl(l)$$

It is also formed when chlorine is passed into ice-cold concentrated hydrochloric acid and lead(II) chloride.

$$PbCl_2(s) + Cl_2(g) + 2Cl^-(aq) \longrightarrow PbCl_6^{2-}(aq)$$

Properties

Oily, yellow, covalent liquid showing similarities to silicon(IV) chloride. It is rapidly hydrolysed by water.

$$PbCl_4(l) + 2H_2O(l) \longrightarrow PbO_2(s) + 4HCl(aq)$$

Lead(II) compounds

Lead(II) oxide, PbO

Preparation A white gelatinous precipitate is formed when ammonia solution or aqueous alkali is added to a solution of lead nitrate. This lead(II) hydroxide may be a hydrated oxide.

$$Pb^{2+}(aq) + 2OH^-(aq) \longrightarrow Pb(OH)_2(s)$$

On heating the solid it forms lead(II) oxide.

Properties

Dimorphic, existing in two crystalline forms: **litharge**, reddish-yellow with a tetragonal lattice and **massicot**, yellow with a rhombic crystal lattice. If lead is carefully heated to just above its melting point, massicot is formed. If massicot is heated to fusion and then ground up, or if lead is strongly heated in a blast of air, litharge is formed. Litharge is in stable form at room temperature but massicot changes to litharge very slowly.

1 Slightly soluble in water.

$$PbO(s) + H_2O(l) \rightleftharpoons Pb^{2+}(aq) + 2OH^-(aq)$$

Because of this, litharge shows amphoteric characteristics.

2 Dissolves easily in acids to give lead(II) salts.

$$PbO(s) + 2HNO_3(aq) \longrightarrow Pb(NO_3)_2(aq) + H_2O \ .$$

3 Dissolves slowly in alkalis to form tetrahydroxoplumbites(II).

$$PbO(s) + 2OH^-(aq) + H_2O(l) \longrightarrow [Pb(OH)_4]^{2-}(aq)$$

4 Is reduced, e.g. by hydrogen, carbon and carbon monoxide.

Uses
In the manufacture of optical glass, as a drier for paints and varnish, and in lead accumulators.

Dilead(II) lead(IV) oxide, Pb_3O_4, red lead oxide
Manufacture Lead(II) oxide (massicot not litharge) heated in air for some hours at 700 K. Oxide is black when hot and scarlet when cool.

$$6PbO(s) + O_2(g) \rightleftharpoons 2Pb_3O_4(s)$$

If massicot is heated to a higher temperature, litharge is formed and this does not readily convert to red lead oxide again.

Properties
Probably not a true oxide and behaves as a mixture of lead(II) oxide and lead(IV) oxide, $2PbO.PbO_2$, suggesting it is the salt of orthoplumbic acid H_4PbO_4. With dilute nitric acid, Pb^{2+} dissolves while PbO_2 precipitates.

$$Pb_3O_4(s) + 4HNO_3(aq) \longrightarrow 2Pb(NO_3)_2(aq) + PbO_2(s) + 2H_2O(l)$$

Uses
In making glazes, optical glass, connecting joints in pipes and as an oxidizing agent in matches and in lead accumulators.

Lead(II) chloride, $PbCl_2$. Precipitated when **dilute hydrochloric acid is** added to **lead(II) nitrate** solution.

$$Pb(NO_3)_2(aq) + 2HCl(aq) \longrightarrow PbCl_2(s) + 2HNO_3$$

It is sparingly soluble in cold water and more soluble in hot water. In concentrated hydrochloric acid, a soluble complex is formed.

$$PbCl_2(s) + 2Cl^-(aq) \rightleftharpoons PbCl_4^{2-}(aq)$$

Lead(II) chloride is essentially ionic. It differs from tin(II) chloride and is not a reducing agent (inert pair effect).

Lead(II) chlorofluoride, PbClF. Salt in which the lattice has lead, chloride and fluoride ions.

Lead(II) sulphate, $PbSO_4$. Precipitated as a white solid by the action of dilute sulphuric acid on a solution of lead(II) nitrate.

$$Pb(NO_3)_2(aq) + H_2SO_4(aq) \longrightarrow PbSO_4(s) + 2HNO_3(aq)$$

Lead(II) carbonate, $PbCO_3$. The normal carbonate is precipitated by adding a solution of sodium hydrogencarbonate to a solution of lead(II) nitrate.

$$Pb^{2+}(aq) + 2HCO_3^-(aq) \longrightarrow PbCO_3(s) + H_2O(l) + CO_2(g)$$

Basic carbonate, $Pb(OH)_2.2PbCO_3$, white lead. Precipitated when sodium carbonate solution is added to a solution of lead(II) nitrate. Used in the paint trade – has good covering qualities but poisonous to children and blackens due to the formation of black lead(II) sulphide.

Lead(II) iodide, PbI_2. Formed by mixing cold aqueous solutions of lead(II) nitrate and potassium iodide.

$$Pb(NO_3)_2(aq) + 2KI(aq) \longrightarrow PbI_2(s) + 2KNO_3(aq)$$

Dissolves in hot water to form a colourless solution, crystallizes on cooling as yellow spangles.

Lead(II) chromate, $PbCrO_4$. Obtained as a bright yellow solid on mixing solutions of lead(II) nitrate and potassium chromate(VI).

$$Pb(NO_3)_2(aq) + K_2CrO_4(aq) \longrightarrow PbCrO_4(s) + 2KNO_3(aq)$$

Lead nitrate, $Pb(NO_3)_2$. Obtained by the action of excess dilute nitric acid on the oxide, hydroxide or carbonate of lead. The solution is filtered, heated until saturated at room temperature, cooled and lead(II) nitrate crystallized out. The white crystals are washed with a little cold water and dried.

$$PbCO_3(s) + 2HNO_3(aq) \longrightarrow Pb(NO_3)_2(aq) + H_2O(l) + CO_2(g)$$

Less soluble than most nitrates and crystals are anhydrous so that it is used to prepare nitrogen dioxide.

$$2Pb(NO_3)_2(s) \longrightarrow 2PbO(s) + 4NO_2(g) + O_2(g)$$

Lead(II) sulphide, PbS, galena.   Obtained in the laboratory as a black solid by passing hydrogen sulphide through a hot solution of lead(II) chloride in hydrochloric acid. Differs from tin(II) sulphide in not being soluble in ammonium sulphide solution. Formed in paints in the presence of atmospheric pollution. Pictures are treated with hydrogen peroxide which forms white lead(II) sulphate.

$$PbS(s) + H_2O_2(aq) \longrightarrow PbSO_4(s) + 4H_2O(l)$$

Lead accumulator

Two lead plates are immersed in an electrolyte of sulphuric acid. The lead plates are usually perforated and one is packed with lead(IV) oxide and the other with spongy lead. An inert porous insulator acts as a separator between the plates.

   To charge: a low direct current is passed through the cell.

Cathode

$$Pb^{2+}(aq) + 2e^- \longrightarrow Pb(s)$$
$$SO_4^{2-} \qquad\qquad \longrightarrow \text{insoluble}$$

Anode

$$Pb + H_2O(l) - 2e^- \longrightarrow PbO(s) + 2H^+(aq)$$
$$SO_4^{2-} \qquad\qquad \longrightarrow \text{into solution}$$

During charging, the cathode acquires a deposit of spongy lead and the anode one of lead(IV) oxide. The passage of hydrogen and sulphate ions into solution increases the concentration of sulphuric acid.

   To discharge

Cathode

$$Pb(s) \qquad\qquad \longrightarrow Pb^{2+} + 2e^-$$
$$Pb^{2+} + SO_4^{2-} \longrightarrow PbSO_4(s)$$

Anode

$$PbSO_4 \text{ deposited}$$

Practice questions

1   State the oxidation states of carbon, silicon, germanium, tin and lead. Relate these to their electronic configurations and outline how the valencies change down the group.

2   Compare the chemistry of carbon and silicon. How are the fundamental differences explained by the electronic configurations of the two elements?

3 What is allotropy? Illustrate your answer with reference to carbon and tin.

4 Write brief notes on the chemistry of (a) the hydrides, (b) the oxides, (c) the chlorides of the group IV elements, carbon→ lead.

5 Explain clearly why the chemistry of carbon is unique.

6 Draw a diagram to show the electronic structure of carbon monoxide. How is the gas prepared in the laboratory?

7 How does carbon monoxide react with (a) sodium hydroxide, (b) copper(II) oxide, (c) iodine(V) oxide?

8 How does carbon monoxide react (a) nickel, (b) chlorine, (c) sulphur?

9 What is (a) producer gas, (b) water gas? How are they made and why are they important?

10 How is carbon dioxide prepared in the laboratory?

11 Outline the reaction between carbon dioxide and (a) water, (b) sodium hydroxide.

12 Give equations for the action of heat on the carbonates of (a) potassium, (b) calcium.

13 Suggest reasons why aluminium carbonate is not precipitated when sodium carbonate solution is added to a solution of an aluminium salt.

14 How is carbon disulphide obtained commercially? How does it react with (a) oxygen, (b) sodium sulphide?

15 Write notes on (a) the hydrides, (b) the halides of silicon. How do these differ from the hydrides and halides of carbon?

16 Write notes on (a) silica, (b) silicates.

17 Compare the chemistry of silicon and boron.

18 How is tin extracted?

19 Discuss allotropy in group IV elements.

20 How does tin react with (a) air, (b) acids, (c) caustic alkalis?

21 State and describe the chlorides of tin.

22 Compare the chemistry of tin(II) and tin(IV).

23 How is lead extracted?

24 Explain why lead is not suitable as piping for soft water.

25 Compare the chemistry of lead(II) and lead(IV).

26 Outline the reaction between dilead(II), lead(IV) oxide and dilute nitric acid.

27 Outline the preparation of (a) lead(IV) chloride, (b) lead(II) carbonate and (c) lead(II) nitrate.

9   p-BLOCK ELEMENTS OF GROUP V

Nitrogen, phosphorus, arsenic, antimony and bismuth have an outer electronic configuration of ns^2np^3 and principal oxidation states of $+3$ and $+5$.

3	N	$1s^2$ $2s^2$	$2p^3$					
3 5	P	$1s^2$ $2s^2$	$2p^6$ $3s^2$	$3p^3$				
3 5	As	$1s^2$ $2s^2$	$2p^6$ $3s^2$	$3p^{10}$ $4s^2$	$4p^3$			
3 5	Sb	krypton core	$4d^{10}$	$5s^2$	$5p^3$			
3 5	Bi		$4d^{10}$ $4f^{14}$ $5s^2$	$5p^6$ $5d^{10}$ $6s^2$	$6p^3$			

General properties

1 With the exception of nitrogen, which has no d-orbitals, the elements attain a maximum covalency of 5 by using all outer 5 electrons to form bonds. The s-electrons are unpaired and promoted to the d-level.

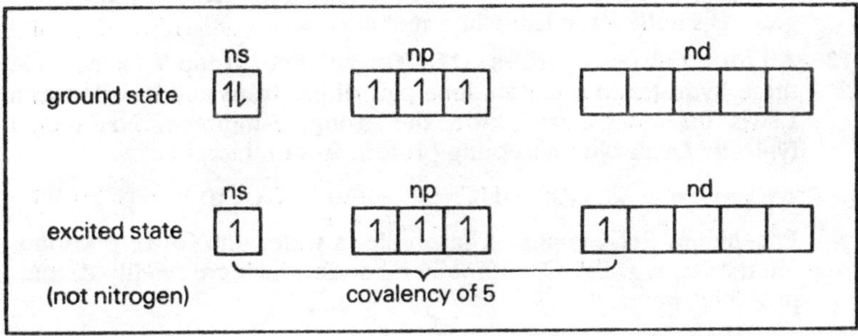

All five covalent bonds are identical, sp^3d hybridization.

2 Down the group, the tendency for the s-electrons to remain as an inert pair increases and the stability of the valency 3 increases.

3 The valency of bismuth is normally 3 and Bi($+5$) has strong oxidizing properties. Thus, sodium bismuthate(V), $NaBiO_3$, can change Mn($+2$) to Mn($+7$) in acidic solution and Bi($+5$) is reduced to Bi($+3$).

4 Bonding is mainly covalent, especially for the higher oxidation states.

5 The formation of positive ions is difficult because of the high ionization energy needed, even Bi^{3+} is not very stable.

6 The high energy needed to acquire three electrons, forming the anion A^{3-}, means only nitrogen has any appreciable tendency to form such an anion. The electronegativity of nitrogen is important

in its ability to form hydrogen bonds. Only fluorine and oxygen are more electronegative than nitrogen.

7 The elements react with the electropositive elements of groups I and II to form metallic nitrides, phosphides, antimonides and bismuthides which generally have giant molecular structures with some ionic character.

8 The transition from non-metallic to metallic character on descending the group is slower than in groups III and IV: nitrogen and phosphorus are non-metals, arsenic and antimony have some metallic properties while bismuth is almost as metallic as lead.

9 Nitrogen is gaseous and diatomic and because of the very high bond enthalpy the molecule is very stable. The other elements in the group can form chains of atoms of different lengths, e.g. P_4.

10 Below 973 K phosphorus vapour consists of P_4 molecules, above this P_2 and at higher temperatures the vapour consists of single atoms.

11 **Allotropy** Phosphorus – white and red, also black and violet. Arsenic – yellow, black and grey forms. Antimony – yellow and grey. Bismuth – reddish-white metal.

12 All form gaseous hydrides AH_3. The ability of group V elements in these hydrides to donate a lone pair of electrons and function as a Lewis base decreases down the group. Ammonia, NH_3, is a typically weak base accepting protons from mineral acids.

$$NH_3(g) + HCl(g) \longrightarrow NH_4Cl(s)$$

Phosphine, PH_3, is almost insoluble in water and forms phosphonium salts, e.g. PH_4Cl with mineral acids which are readily decomposed by water.

$$PH_4^+ + (aq) + H_2O(l) \longrightarrow PH_3(g) + H_3O^+(aq)$$

13 Thermal stability of hydrides decreases down the group. Ammonia is fairly stable to heat, phosphoric and arsine decompose on heating, stibine and bismuthine are unstable at room temperature.

14 As A—H strength lessens, it is easier for other atoms or groups to replace hydrogen atoms in the respective molecules.

15 Ammonia reduces some metal oxides, e.g. copper(II) oxide at red-heat, phosphine is a fairly strong reducing agent precipitating copper and silver from aqueous solutions of their salts.

$$4Cu^{2+}(aq) + PH_3(g) + 4H_2O(l) \longrightarrow$$
$$4Cu(s) + H_3PO_4(aq) + 8H^+(aq)$$

16 All form halides AX_3 which are covalent (NBr_3 and NI_3 are not known but complexes $NBr_3.6NH_3$ and $NI_3.NH_3$ are known). Ionic character increases down the group. Nitrogen trichloride is an explosive oil, phosphorus trichloride is a colourless fuming liquid, hydrolysed by water, trihalides of arsenic, antimony and bismuth are progressively more ionic. $SbCl_3$ and $BiCl_3$ are partially hydrolysed in water.

17 Nitrogen is not able to form a pentahalide because there are no d-orbitals available.

18 All form more than one oxide, e.g. nitrogen forms five. The oxides of nitrogen differ from those of phosphorus because phosphorus forms polymeric structures with oxygen. Phosphorus trioxide, P_4O_{10} has three crystalline forms, the trioxides of antimony and bismuth have giant molecular structures transitional in character between typically covalent and ionic lattices.
N_2O and NO(neutral); N_2O_3, N_2O_4 and N_2O_3(acidic); P_4O_6 and P_4O_{10}(acidic); As_4O_6 and As_4O_{10}(acidic); Sb_2O_3(amphoteric); Sb_2O_3(acidic); Bi_2O_3(basic); B_2O_5(unstable acidic).

19 The formation of oxides and oxyacids is parallel with that of group IV in that phosphorus forms more oxides than nitrogen.

20 Nitrides may be interstitial. They are often hard and refractory (*c.f.* carbon and hydrogen). A metal may be treated with nitrogen to form a hard tough surface (nitriding).

Nitrogen, N_2

Occurrence The free element forms four-fifths of the atmosphere. It also occurs as Chile saltpetre, $NaNO_3$, and in proteins.

Preparation

1 Air is first passed through sodium hydroxide solution to remove carbon dioxide and then through finely divided copper to remove oxygen.

$$2OH^-(aq) + CO_2(g) \longrightarrow CO_3^{2-}(aq) + H_2O(aq)$$

$$2Cu(s) + O_2(g) \longrightarrow 2CuO(s)$$

Residual gas, atmospheric nitrogen, contains the noble gases and is collected over water.

2 A mixture of equimolar amounts of sodium nitrite and ammonium chloride is heated.

$$NH_4^+(aq) + NO_2^-(aq) \longrightarrow N_2(g) + 2H_2O(l)$$

The mixture is warmed to start the reaction, which then continues without further heating. To obtain pure nitrogen, the gas is passed through concentrated alkali to remove any acidic gas or chlorine, then through concentrated sulphuric acid to remove ammonia and to dry the gas and finally over heated copper to decompose any oxides of nitrogen.

3 Orange ammonium dichromate(VI) decomposes on heating to form green chromium(III) oxide and nitrogen. The reaction is exothermic and vigorous. Heat and light are given out and there is a large increase in volume. It is best mixed with three times its mass of sand to moderate the reaction.

$$(NH_4)_2\,Cr_2O_7(s) \longrightarrow Cr_2O_3(s) + 4H_2O(g) + N_2(g)$$

4 Nitrogen is obtained when ammonia is passed over heated copper oxide.

$$3CuO(s) + 2NH_3(g) \longrightarrow 3Cu(s) + 3H_2O(g) + N_2(g)$$

5 Nitrogen is formed when an oxide of nitrogen is passed over finely divided copper at red heat, e.g.

$$Cu(s) + N_2O(g) \longrightarrow CuO(s) + N_2(g)$$

6 Nitrogen is also formed when chlorine is passed into liquid ammonia.

$$8NH_3(aq) + 3Cl_2(g) \longrightarrow 6NH_4Cl(aq) + N_2(g)$$

Excess chlorine can produce the dangerously explosive nitrogen chloride.

$$NH_4Cl(aq) + 3Cl_2(g) \longrightarrow NCl_3(l) + 4HCl(g)$$

Industrial Production
Fractional distillation of liquid air yields oxygen, boiling point 90 K, and nitrogen, boiling point 77 K.
Uses The most important uses are the production of ammonia for fertilizers and as an inert atmosphere.

Physical properties
Colourless, odourless gas, slightly soluble in water.

Chemical properties
1 Nitrogen is relatively inert. Molecules probably have to dissociate before a reaction occurs and nitrogen is stable unless heated.
2 The nitrogen atom cannot have more than 8 electrons in its outer electron level so that it can form a maximum of 4 covalent bonds, e.g. in NH_4^+. The bonds are identical because of sp^3 hybridization.
3 With metals, at red-heat, nitrides may be formed, e.g.

$$3Mg(s) + N_2(g) \longrightarrow Mg_3N_2(s)$$
$$3Ca(s) + N_2(g) \longrightarrow Ca_3N_2(s)$$
$$2Al(s) + N_2(g) \longrightarrow 2AlN(s)$$

Nitrides form ammonia with water, e.g.

$$Mg_3N_2(s) + 6H_2O(l) \longrightarrow 3Mg(OH)_2(s) + 2NH_3(g)$$

4   With non-metals, nitrides are formed, e.g.

$$2B(s) + N_2(g) \longrightarrow 2BN(s)$$
$$6Si(s) + 4N_2(g) \longrightarrow 2Si_3N_4(s)$$

5   Nitrogen forms several oxides. Above 1300 K, it combines with oxygen to form nitrogen oxide (nitric oxide).

$$O_2(g) + N_2(g) \longrightarrow 2NO(g)$$

6   At red-heat, under 2 atmospheres pressure, nitrogen combines with calcium dicarbide to form calcyanamide which is used as a fertilizer (nitrolin).

$$CaC_2(s) + N_2(g) \longrightarrow CaCN_2(s) + C(s)$$

Nitrogen Cycle

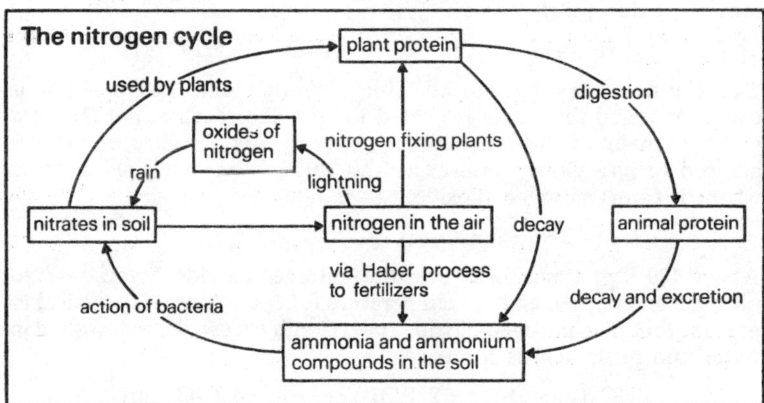

Fixation of Nitrogen

1   Cyanamide process in which nitrogen from the air is converted to calcium cyanamide.

2   **The Haber Process** in which nitrogen and hydrogen combine to form ammonia.

$$3H_2(g) + N_2(g) \rightleftharpoons 2NH_3(g)$$

Applying Le Chatelier's principle to this equation.

(i)   **Pressure**   The formation of ammonia is accompanied by a decrease in volume and is therefore favoured by a high pressure. $2 \times 10^7 \, N\,m^{-2}$ (200 atm.) is used in some plants but it may be as high as $10^8 \, N\,m^{-2}$ (1000 atm.)

(ii)   **Temperature**   The formation of ammonia is exothermic and is favoured by a low temperature, but this would slow down the reaction. The optimum temperature is about 720 K.

(iii) **Catalyst**   Finely divided iron promoted with alkali or alumina.

(iv)   **Equilibrium**   This is not reached because it takes too long. The ammonia is removed and the unchanged gases re-circulated.

(v)   The **nitrogen** is obtained by the fractional distillation of liquid air.

(vi)   The **hydrogen** is obtained by the steam reforming of methane in natural gas. Gaseous naphtha from the refining of petroleum may also be used (see section on hydrogen).

(vii) The ammonia obtained is dissolved in water or changed to nitric acid or ammonium sulphate.

Production of Nitric Acid from Ammonia

A mixture of air (90%) and ammonia (10%) is passed through a catalyst of platinum gauze at red-heat (1120 K). The ammonia is oxidized to nitrogen oxide (nitric oxide).

$$4NH_3(g) + 5O_2(g) \longrightarrow 4NO(g) + 6H_2O(g)$$

The platinum may be alloyed with 10% rhodium. The reaction is exothermic and the gauze is heated to start the reaction but the heat given out maintains the temperature. After cooling, the nitrogen oxide is allowed to pass slowly with excess air through an oxidizing chamber where it forms nitrogen dioxide.

$$2NO(g) + O_2(g) \rightleftharpoons 2NO_2(g)$$

Above 420 K at atmospheric pressure, nitrogen dioxide dissociates into nitrogen and oxygen and the temperature is kept as low as is practical to prevent this. The nitrogen dioxide, mixed with excess air, is absorbed in water and nitric acid is formed.

$$4NO_2(g) + O_2(g) + 2H_2O(l) \longrightarrow 4HNO_3(aq)$$

The temperature is kept above 350 K to prevent the formation of nitrous acid, which is unstable in heat. A pressure of $7 \times 10^5\,N\,m^{-2}$ (7 atm.) is maintained to keep the gases flowing through the plant.

Ammonia, NH₃

Preparation   Ammonia is obtained by heating a mixture of alkali and an ammonium salt, usually calcium hydroxide and ammonium chloride.

$$2NH_4Cl(s) + Ca(OH)_2(s) \longrightarrow CaCl_2(s) + 2H_2O(g) + 2NH_3(g)$$

The flask is tilted to prevent the water formed trickling back on to the hot surface and cracking the glass. The ammonia is dried with lumps of quicklime and collected by upward delivery, because it is less dense than air.

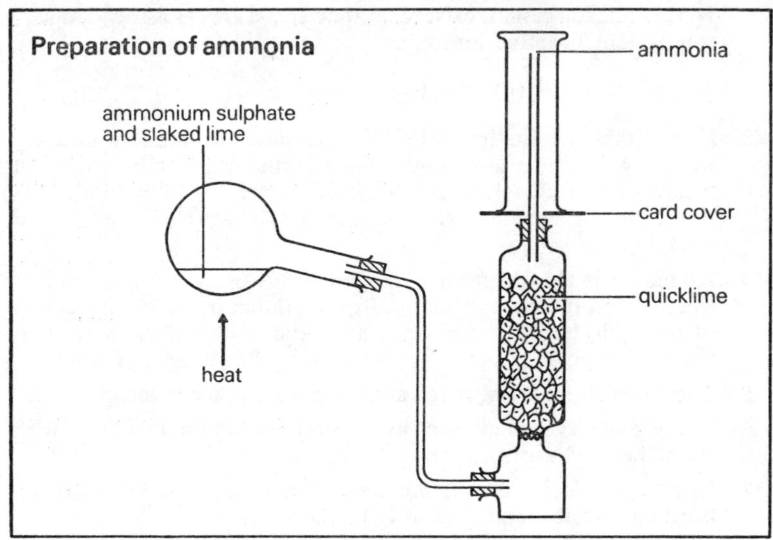

Production

1 **Haber Process** – see beginning of this section.
2 **Fractional distillation of the ammoniacal liquor obtained by the destructive distillation of coal** for coal gas. Usually the ammonia is changed to ammonium sulphate and used as a fertilizer.

Physical properties

1 Colourless gas with a pungent choking smell.
2 It has a high critical temperature and is easily liquefied at room temperature by pressure.
3 In large or sudden doses, ammonia may cause death through the paralysis of the muscles connected with respiration or through vagal reflex.
4 **Very soluble** in water – shown by the fountain experiment.

Chemical properties

1 The solution formed by ammonia in water is weakly ionized. It is an electrolyte, turns red litmus blue and precipitates metal hydroxides, e.g.

$$Fe^{3+}(aq) + 3OH^-(aq) \longrightarrow Fe(OH)_3(s)$$

$$Al^{3+}(aq) + 3OH^-(aq) \longrightarrow Al(OH)_3(s)$$

2 The ammonium ion is formed when the lone pair of electrons on the nitrogen atom is used to make a dative covalent bond with a proton,

NH_4^+. The ion cannot exist separately and is always associated with a univalent negative ion, e.g.

$$NH_3(g) + HCl(g) \longrightarrow NH_4^+Cl^-(s)$$

3   Forms stable ammonium salts with all mineral acids and with many weaker acids. The ammonium ion is actually a complex ion but behaves like a univalent metallic ion and NH_4^+ is isoelectronic with Na^+ as in, for example, Na^+Cl^- and $NH_4^+Cl^-$; $(Na^+)_2 SO_4^{2-}$ and $(NH_4^+)_2SO_4^{2-}$.

4   Ammonia is able to form hydrogen bonds using the lone pair of electrons on the nitrogen atom. Consideration of the boiling points of the hydrides of phosphorus, arsenic and antimony shows that the boiling point of ammonia is raised by this hydrogen bonding.

5   The lone pair on the nitrogen atom creates a dipole in the molecule.

6   Because of the formation of hydrogen bonds, ammonia has a high latent heat of vaporization.

7   In pure liquid ammonia, the ionization is reduced by hydrogen bonding and the equilibrium is to the left.

$$c.f. \quad \begin{bmatrix} 2NH_3 & \rightleftharpoons & NH_4^+ + NH_2^- \\ 2H_2O & \rightleftharpoons & H_3O^+ + OH^- \end{bmatrix}$$

8   Ionization of the aqueous solution of ammonia is reduced by hydrogen bonding.

$$\begin{array}{c} H \\ | \\ H-N\cdots\cdots H-O-H \\ | \\ H \end{array}$$

In $NH_3(aq) + H_2O(l) \rightleftharpoons NH_4^+(aq) + OH^-(aq)$ the equilibrium lies to the left. If all the hydrogen atoms of NH_4^+ are replaced by organic groups such as methyl, CH_3, no hydrogen bonding is possible and ionization is almost complete so that $N(CH_3)_4^+ OH^-$, tetramethyl-ammoninium hydroxide, is a very strong base.

9   In liquid ammonia, ammonium salts behave as strong acids and amides behave as strong bases and so ammonium chloride will react with potassium amide to form a salt and ammonia.

$$2NH_3 \rightleftharpoons \underset{\substack{\text{Lewis} \\ \text{acid}}}{NH_4^+} + \underset{\substack{\text{Lewis} \\ \text{base}}}{NH_2^-}$$

$$\therefore \quad \underset{\text{acid}}{NH_4Cl} + \underset{\text{base}}{KNH_2} \longrightarrow \underset{\text{salt}}{KCl \downarrow} + \underset{\text{solvent}}{2NH_3}$$

10   Complexes   The ammonia molecule uses the lone pair of electrons on the nitrogen to form a dative covalent bond with some metallic ions, e.g. $[Cu(NH_3)_4]^{2+}$ and $[Co(NH_3)_6]^{3+}$.

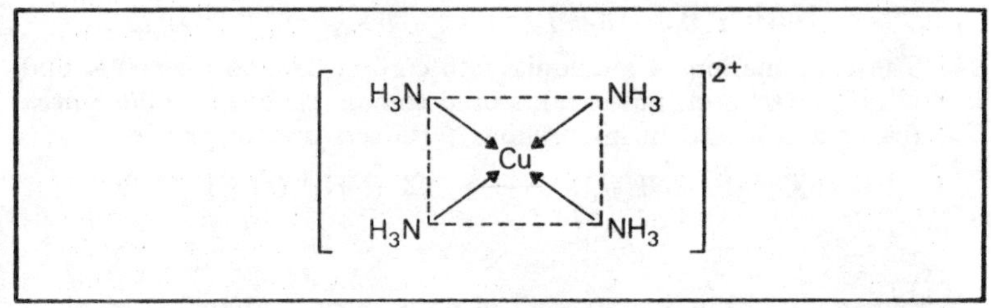

11   The ammonia molecule is pyramidal and the ammonium ion is tetrahedral with four equivalent covalent N—H bonds, a hydrogen atom occupying each corner.

12   Oxidation of ammonia   The products depend on the conditions and reactants.

(i)   Ammonia will not burn in air but if there is more than 25% oxygen present, ammonia may burn explosively if the mixture is sparked forming water and nitrogen.

$$4NH_3(g) + 3O_2(g) \longrightarrow 6H_2O(g) + 2N_2(g)$$

(ii)   When ammonia mixed with excess air is passed over red-hot platinum as a catalyst, nitrogen oxide and steam are formed.

$$4NH_3(g) + 5O_2(g) \longrightarrow 4NO(g) + 6H_2O(g)$$

(iii)   Ammonia will reduce some metal oxides, e.g. when passed over heated copper(II) oxide, the products are copper, nitrogen and steam.

$$3CuO(s) + 2NH_3(g) \longrightarrow 3Cu(s) + 3H_2O(g) + N_2(g)$$

(iv)   Ammonia is readily oxidized by chlorine. When concentrated ammonia is allowed to drop into chlorine gas, there is a greenish flame and dense white fumes of ammonium chloride are formed. Dangerously explosive nitrogen trichloride may also be formed.

Excess ammonia:

$$8NH_3(g) + 3Cl_2(g) \longrightarrow 6NH_4Cl(s) + N_2(g)$$

Excess chlorine:

$$NH_3(g) + 3Cl_2(g) \longrightarrow NCl_3(l) + 3HCl(g)$$

13   With metals, ammonia may form amides, e.g. dry ammonia reacts with sodium at 670 K to form sodamide. This forms ammonia with water.

$$2Na(l) + 2NH_3(g) \longrightarrow 2NaNH_2(s) + H_2(g)$$

$$NaNH_2(s) + H_2O(l) \longrightarrow NaOH(aq) + NH_3(g)$$

14   The combination of ammonia with carbon dioxide (at 460 K and N m^{-2}, 1000 atm. pressure) is used commercially to produce **urea** (carbamide) used in nitrogenous fertilizers and in plastics.

$$CO_2(g) + 2NH_3(g) \longrightarrow CO(NH_2)_2(s) + H_2O(l)$$

Uses
1   Industrially in the manufacture of urea, nitric acid and ammonium sulphate.
2   Used in the Solvay process for the manufacture of sodium carbonate.
3   Liquid ammonia used to be used as a refrigerant because of its volatility and high molar enthalpy of vaporization. Replaced in domestic refrigerators because it is toxic and corrosive.
4   Used in the manufacture of some man-made fibres such as nylon.
5   Used in the laundering of woollens, in cleaning polishes for silver and in smelling salts.

Ammonium compounds

1   All decompose on heating and some sublime, e.g.

$$NH_4Cl(s) \rightleftharpoons NH_4(g) + HCl(g)$$

$$NH_4NO_3(s) \longrightarrow N_2O(g) + 2H_2O(g)$$

2   Form ammonia when heated with alkali

$$NH_4^+ + OH^- \longrightarrow NH_3(g) + H_2O(g)$$

3   **Nessler's Reagent**   Potassium iodide is added to mercury(II) chloride until the scarlet precipitate of mercury(II) iodide formed is just redissolved and then potassium hydroxide solution is added until the liquid is alkali. Nessler's reagent is a solution of potassium tetraiodomercury(II). When added in excess to a solution containing ammonium ions, a yellowish-brown, brown precipitate is formed $(OHg_2)NH_2I$.

4   **Ammonium chloride**, NH_4Cl
Formed by adding dilute hydrochloric acid to ammonia until the solution is just acidic

$$NH_3(aq) + HCl(aq) \longrightarrow NH_4Cl(aq)$$

On a large scale, a mixture of ammonium sulphate and sodium chloride is heated when ammonium chloride sublimes.

$$(NH_4)_2SO_4(s) + 2NaCl(s) \longrightarrow Na_2SO_4(s) + 2NH_4Cl(s)$$

Ammonium chloride sublimes at 620 K and just above this temperature is approximately 85% dissociated. It is used in the Leclanché battery and as a flux for soldering.

5 Ammonium sulphate, $(NH_4)_2SO_4$
This is used as a nitrogenous fertilizer.

6 Ammonium nitrate, NH_4NO_3
On a large scale, ammonia gas reacts with concentrated nitric acid to form fused nitrate.

$$NH_3(g) + HNO_3(conc.aq.) \longrightarrow NH_4NO_3(s)$$

Alternatively concentrated solutions of ammonium sulphate and calcium nitrate are mixed, when the sparingly soluble calcium sulphate is precipitated.

$$(NH_4)_2SO_4(aq) + Ca(NO_3)_2(aq) \longrightarrow$$
$$2NH_4NO_3(aq) + CaSO_4(s)$$

Used as a fertilizer Ammonium nitrate is so soluble it would be washed out of the soil so droplets of molten ammonium nitrate are allowed to fall down a tower with a little water forming glassy granules which dissolve slowly in the soil. **Amatol** (explosive) contains 80% ammonium nitrate and 20% T.N.T. Ammonium nitrate dissolves in water with the absorption of heat – **a freezing mixture.**

7 Ammonium carbonate, $(NH_4)_2CO_3$
Obtained by heating a mixture of calcium carbonate and ammonium sulphate when ammonium carbonate sublimes and then re-subliming the compound in the presence of a little water. The main substance is actually ammonium carbamate, $(NH_2COO)NH_4$, together with a little ammonium hydrogencarbonate. In solution, ammonium carbonate is present.

$$NH_2COO^-(aq) + NH_4^+(aq) + H_2O(l) \rightleftharpoons$$
$$2NH_4^+(aq) + CO_3^{2-}(aq)$$

8 Ammonium sulphide
When hydrogen sulphide is passed through ammonia solution, the hydrogen sulphide NH_4HS and a normal sulphide are formed on standing. On exposure to air, the liquid becomes yellow as polysulphides are formed.

Hydrazine, N_2H_4 $H_2N—NH_2$
1 The hydride of N_2.

2 Colourless liquid, poisonous and boils with decomposition. Soluble in water and alcohol.

3 Hydrazine hydrate is used as a rocket fuel.

4 In aqueous solution, hydrazine acts as a weak di-acid base forming salts, e.g. N_2H_4HCl and N_2H_42HCl.

5 Hydrazine and its sulphate are strong reducing agents, reducing iron(III) to iron(II), manganate(VII) to manganate(II) and iodine (in KI) to iodide ions, e.g.

$$4Fe^{3+}(aq) + N_2H_4(aq) \longrightarrow 4Fe^{2+}(aq) + 4H^+(aq) + N_2(g)$$

6 Organic compounds containing the carbonyl group $\diagdown C{=}O$ (aldehydes and ketones) form condensation compounds – hydrazones, e.g.

$$\underset{\displaystyle CH_3-\overset{\displaystyle\overset{H}{|}}{C}=\boxed{O + H_2}N-NH_2}{} \longrightarrow \underset{\displaystyle CH_3-\overset{\displaystyle\overset{H}{|}}{C}=N-NH_2 + H_2O}{}$$

Hydroxylamine, NH_2OH

1 Colourless solid usually used as its salts. Decomposes explosively on heating.

$$3NH_2OH(s) \longrightarrow NH_3(g) + 3H_2O(l) + N_2(g)$$

2 Readily soluble in water.

3 Acts as a weak base forming $NH_3OH^+Cl^-$ and $(NH_3OH^+)_2SO_4^{2-}$, both being hydrolysed in solution.

4 Acts as a reducing agent, e.g. copper(II) salts in the presence of alkali (Fehling's solution) to copper(II) oxide, and mercury(II) chloride to mercury(I) chloride.

$$2NH_2OH(aq) + 4Cu^{2+}(aq) + 8OH^-(aq) \longrightarrow$$
$$N_2O(g) + 2Cu_2O(s) + 7H_2O(l)$$

5 In the presence of alkali, iron(II) is oxidized to iron(III), e.g. green iron(II) hydroxide to brown iron(III) hydroxide.

$$2Fe(OH)_2(s) + NH_2OH(aq) + H_2O(l) \longrightarrow$$
$$2Fe(OH)_3(s) + NH_3(g)$$

In the presence of acid, iron(III) is reduced to iron(II).

$$4Fe^{3+}(aq) + 2NH_2OH(aq) \longrightarrow$$
$$N_2O(g) + H_2O(l) + 4H^+(aq) + 4Fe^{2+}(aq)$$

6 Condensation occurs with carbonyl compounds forming oximes.

$$\underset{\displaystyle CH_3-\overset{\displaystyle\overset{H}{|}}{C}=\boxed{O + H_2}NOH}{} \longrightarrow \underset{\displaystyle CH_3-\overset{\displaystyle\overset{H}{|}}{C}=NOH + H_2O}{}$$

Oxides and oxy-acids of nitrogen

1 Dinitrogen oxide (nitrous oxide), N_2O (neutral).
2 Nitrogen oxide (nitric acid), NO (neutral).
3 Dinitrogen trioxide, N_2O_3 (known but not stable at room temperature), (acidic).
4 Nitrogen dioxide (dinitrogen tetroxide), N_2O_4 (acidic).
5 Dinitrogen pentoxide, N_2O_5 (acidic).

Dinitrogen oxide, N_2O

Preparation   Equimolar quantities of potassium nitrate and ammonium chloride are mixed and heated, this is safer than heating ammonium nitrate alone because the last small amount may decompose explosively.

$$NH_4Cl(s) + KNO_3(s) \longrightarrow KCl(s) + 2H_2O(l) + N_2O(g)$$

The gas is collected over warm water. To obtain pure, the gas is passed through iron(II) sulphate to remove nitrogen oxide and through caustic alkali to remove chlorine.

Properties

1 Colourless gas with a sweetish smell.
2 Anaesthetic used in dentistry – laughing gas.
3 Fairly soluble in cold water giving a neutral solution.
4 Dinitrogen oxide is formed when nitric(I) acid (hyponitrous acid) is heated but the reaction is not reversible.

$$H_2N_2O_2(s) \longrightarrow H_2O(l) + N_2O(g)$$

5 **Combustion** – if the burning material is hot enough to decompose nitrogen dioxide, the resulting mixture contains more oxygen than air and the material continues to burn.

$$2N_2O(g) \longrightarrow 2N_2 + O_2(g)$$
$$33\% \text{ by vol.}$$

(i)   **Magnesium** continues to burn with a bluish-white flame.

$$Mg(s) + N_2O(g) \longrightarrow MgO(s) + N_2(g)$$

(ii)   **Phosphorus** burns with a bright yellow flame forming phosphorus(V) oxide.

$$P_4(s) + 10N_2O(g) \longrightarrow P_4O_{10}(s) + 10N_2(g)$$

(iii)   **Sulphur**, if burning vigorously, continues to burn with a bright purple flame forming sulphur dioxide.

$$S(s) + 2N_2O(g) \longrightarrow SO_2(g) + 2N_2(g)$$

(iv)   A **candle** continues to burn forming carbon dioxide and water vapour.

(v)   A **glowing splint** is rekindled by nitrogen dioxide – the gas is easily decomposed at this relatively low temperature.

(vi) **Copper** at red-heat decomposes the gas and copper(II) oxide and nitrogen are formed.

$$Cu(s) + N_2O(g) \longrightarrow CuO(s) + N_2(g)$$

6 Dinitrogen oxide differs from oxygen in that it does not react with nitrogen oxide while oxygen does react to form brown fumes of nitrogen dioxide.

7 Structure A linear molecule which is probably a resonance hybrid of two forms.

$$^+N\!=\!N^-\!=\!O \longleftrightarrow N\!\equiv\!N^+\!-\!O^-$$

Nitrogen oxide, NO

Preparation Formed when concentrated nitric acid, diluted with an equal volume of water, is added to copper turnings. The reaction occurs in the cold and the reaction flask is filled with brown fumes as nitrogen oxide reacts with the air in the flask to form nitrogen dioxide.

$$3Cu(s) + 8HNO_3(aq) \longrightarrow 3Cu(NO_3)_2(aq) + 4H_2O(l) + 2NO(g)$$

$$2NO(g) + O_2(g) \rightleftharpoons 2NO_2(g)$$

The gas is collected over water and the brown fumes dissolve. As the nitric acid becomes more dilute, dinitrogen oxide and nitrogen may be formed. Nitrogen oxide is purified by absorption in cold iron(II) sulphate, from which it is expelled on heating.

Properties

1 Colourless gas which forms brown fumes with air. Its smell is unknown.

$$2NO(g) + O_2(g) \rightleftharpoons 2NO_2(g)$$

2 Slightly soluble in water.

3 Gas is more stable than dinitrogen oxide and does not dissociate until 1300 K.

4 Burning materials are extinguished unless the flames are hot enough to break up the dinitrogen oxide. Thus, a glowing splint, a candle and sulphur are extinguished.

(i) **Magnesium** continues to burn with a bluish-white flame.

$$2Mg(s) + 2NO(g) \longrightarrow 2MgO(s) + N_2(g)$$

(ii) **Phosphorus**, if weakly burning, is extinguished but if strongly alight, continues burning with a very bright, yellow flame forming phosphorus(V) oxide.

$$P_4(s) + 10NO(g) \longrightarrow P_4O_{10}(s) + 5N_2(g)$$

(iii) A mixture of **carbon disulphide** vapour and nitrogen oxide burns if ignited.

(iv) **Copper**, when hot, decomposes nitrogen oxide.

$$2Cu(s) + 2NO(g) \longrightarrow 2CuO(s) + N_2(g)$$

5 Addition reactions

(i) At room temperature, nitrogen oxide reacts with oxygen to form nitrogen dioxide.

$$2NO(g) + O_2(g) \longrightarrow 2NO_2(g)$$

(ii) On standing at room temperature, chlorine slowly combines with nitrogen oxide to form nitrosyl chloride; reaction catalysed by divided charcoal.

$$2NO(g) + Cl_2(g) \longrightarrow 2NOCl(g)$$

(iii) Iron(II) sulphate solution at room temperature forms a brown addition compound which decomposes on heating. This is the basis of the brown ring test for a nitrate. In aqueous solution, iron(II) is usually surrounded by six water molecules as ligands. If one of these water molecules is replaced by a nitrogen oxide molecule, the unstable complex ion $[Fe(H_2O)_5NO]^{2+}$ is formed.

6 Oxidation of nitrogen oxide

(i) With oxygen, nitrogen dioxide, NO_2, is formed.
(ii) With iodine in potassium iodide,

$$2NO + 3I_2(aq) + 4H_2O(aq) \longrightarrow$$
$$2NO_3^-(aq) + 8H^+(aq) + 6I^-(aq)$$

(iii) Acidified potassium manganate(VII) is reduced to potassium manganate(II),

$$3MnO_4^-(aq) + 5NO(aq) + 4H^+(aq) \longrightarrow$$
$$3Mn^{2+}(aq) + 5NO_3^-(aq) + 2H_2O(l)$$

7 Reduction of nitrogen oxide

(i) When sparked with hydrogen, nitrogen is formed.

$$2NO(g) + H_2(g) \longrightarrow N_2(g) + 2H_2O(l)$$

(ii) When hydrogen and nitrogen oxide are passed over heated platinum black, ammonia is formed.

$$2NO(g) + 5H_2(g) \longrightarrow 2NH_3(g) + 2H_2O(l)$$

(iii) Forms dinitrogen oxide with sulphur dioxide in the presence of water.

$$2NO(g) + SO_2(g) + H_2O(l) \longrightarrow N_2O(g) + H_2SO_4(aq)$$

8 The nitrogen oxide molecule contains 11 valency electrons, **one of which is unpaired**. This is very rare, the other common example being nitrogen dioxide. Nitrogen oxide is **paramagnetic** and effectively contains a 3-electron bond $\overset{\times}{\underset{\times}{N}}\overset{\times\times\times}{=\!=\!=}\overset{\times}{O}\underset{\times}{}$. It is probably a resonance hybrid.

$$\overset{\times}{\underset{\times}{N}}\overset{\times}{=\!=\!=}\overset{\times}{\underset{\times}{O}} \quad \longleftrightarrow \quad \overset{\times\times}{N}\overset{}{=\!=\!=}\overset{\times}{\underset{\times}{O}}$$

It dimerizes in the solid state to N_2O_4 and this is no longer paramagnetic so that pairing of the odd electron has occurred.

(i) Nitrogen oxide can lose one electron to give the nitrosyl cation (nitrosonium ion) NO^+ forming salts such as $(NO)^+(BF_4)^-$, nitrosyl (tetrafluoroborate).

(ii) Nitrogen oxide can donate electrons as in $[Fe(H_2O)_5NO]^{2+}$ when it replaces a water molecule. Similarly, it can replace a cyanide ion in $[Fe(CN)_6]^{3-}$ to give $[Fe(CN)_5NO]^{2-}$. Here the nitric acid donates its odd electron to the iron and then further donates an electron pair to the metal, *c.f.* carbon monoxide in carbonyls;

$$M^- \longleftarrow \overset{+}{N}\equiv O \quad \text{and} \quad M \longleftarrow C\equiv O$$

where M is the metal.

(iii) Nitrogen oxide can gain an electron, as in the reaction with sodium in liquid ammonia when sodium nitrosyl is formed, Na^+NO^-.

Dinitrogen trioxide, N_2O_3. Is formed as a bluish liquid when a mixture of equal volumes of nitrogen oxide and nitrogen dioxide is cooled to 240 K, possibly as a result of the pairing of the odd electrons in each. The anhydride of nitrous acid formed dissociates completely at room temperature.

$$2HNO_2(aq) \rightleftharpoons H_2O(l) + N_2O_3(aq)$$

$$N_2O_3(g) \rightleftharpoons NO(g) + NO_2(g)$$

Nitrogen dioxide, NO_2 or N_2O_4
Preparation
1 Heat lead(II) nitrate and collect the gas as a dark-green liquid in a U-tube cooled by an ice/salt freezing mixture. The liquid is mainly N_2O_4 and the green colour is caused by the presence of N_2O_3.

$$2Pb(NO_3)_2(s) \longrightarrow PbO(s) + 4NO_2(g) + O_2(g)$$

On cooling,

$$2NO_2(g) \longrightarrow N_2O_4(l)$$

2 Copper reacts with cold concentrated nitric acid to give mainly nitrogen dioxide.

$$Cu(s) + 4HNO_3(conc.) \longrightarrow$$

$$Cu(NO_3)_2(aq) + 2H_2O(l) + 2NO_2(g)$$

3 On heating tin with concentrated nitric acid,

$$Sn(s) + 4HNO_3(conc.) \longrightarrow SnO_2(s) + 2H_2O(l) + 4NO_2(g)$$

Properties

1 Forms a colourless solid, melting at 264 K to a pale yellow liquid.

2 As the temperature rises, the colour of the liquid deepens and just below its boiling point, 295 K, the liquid is orange. On boiling a reddish-brown vapour is formed. At 300 K, the vapour density is 38 and the formula N_2O_4 would require a vapour density of 46. At about 300 K, the gas is approximately 20% dissociated. As the temperature rises, the colour darkens to dark brown and then almost black and at 420 K dissociation is virtually complete to NO_2. With a further rise in temperature, the colour of the gas lightens because thermal dissociation occurs with nitrogen and oxygen being formed. The breakdown is complete at 900 K.

3 Nitrogen dioxide is a mixed acid anhydride forming nitric and nitrous acids with water.

$$2NO_2(g) + H_2O(l) \longrightarrow HNO_3(aq) + HNO_2(aq)$$

4 With caustic alkali nitrites and nitrates are formed.

$$2NO_2(g) + 2OH^-(aq) \longrightarrow NO_2^-(aq) + NO_3^-(aq) + H_2O(l)$$

5 Combustion continues if the flame is hot enough to decompose the nitrogen dioxide. Vigorously burning magnesium and phosphorus continue to burn.

$$4Mg(s) + 2NO_2(g) \longrightarrow 4MgO(s) + N_2(g)$$
$$2P_4(s) + 10NO_2(g) \longrightarrow 2P_4O_{10}(s) + 5N_2(g)$$

6 The gas is decomposed by copper at red-heat.

$$4Cu(s) + 2NO_2(g) \longrightarrow 4CuO(s) + N_2(g)$$

7 Nitrogen dioxide is an odd electron molecule and is paramagnetic. It may be considered as a resonance hybrid of the following:

N_2O_4 is formed by the pairing of the previously unpaired electrons in the NO_2 molecules and is a resonance hybrid of

8 Nitrogen dioxide in the lead chamber process
Nitrogen dioxide oxidizes sulphur dioxide to sulphuric acid in the presence of water.

$$SO_2(g) + H_2O(l) + NO_2(g) \longrightarrow H_2SO_4(l) + NO(g)$$

Nitrogen oxide is an **intermediate compound catalyst** being used up and re-formed as above.

$$NO(g) + \tfrac{1}{2}O_2(g) \longrightarrow NO_2(g)$$

In the absence of water:

$$2NO_2(g) + H_2SO_4(l) \longrightarrow NO.HSO_4(s) + HNO_3(l)$$

Nitrosyl hydrogen sulphate is found as chamber crystals if the water supply is insufficient.

Dinitrogen pentoxide, N_2O_5. Formed when phosphorus(V) oxide is added to ice-cold, freshly made pure nitric acid.

$$4HNO_3(l) + P_4O_{10}(s) \longrightarrow 2N_2O_5(s) + 4HPO_3(s)$$

It is a colourless solid, very hygroscopic and the anhydride of nitric acid.

$$N_2O_5(s) + H_2O(l) \longrightarrow 2HNO_3(l)$$

Structure

$$
\begin{array}{ccc}
O & & O \\
\parallel & & \parallel \\
N-O-N \\
\swarrow & & \searrow \\
O & & O
\end{array}
$$

Oxy-acids

1 Hyponitrous acid (nitric(I) acid), $H_2N_2O_2$.
2 Nitrous acid (nitric(III) acid), HNO_2.
3 Nitric acid (nitric(V) acid), HNO_3.

Hyponitrous acid (nitric(I) acid), $H_2N_2O_2$
Obtained by the action of nitrous acid on hydroxylamine.

$$HNO_2(aq) + NH_2OH(s) \longrightarrow H_2N_2O_2(s) + H_2O(l)$$

A weak dibasic acid. The acid and its salts are reducing agents being oxidized to compounds containing nitrogen in the $+3$ or $+5$ oxidation states.

Nitrous acid (nitric(III) acid), HNO_2
Preparation Obtained by the action of dilute hydrochloric or sulphuric acid on potassium nitrite or sodium nitrite at 273 K.

$$NaNO_2(s) + HCl(aq) \longrightarrow HNO_2(aq) + NaCl(aq)$$

If barium nitrite is used, a calculated quantity of sulphuric acid is used to precipitate barium sulphate, leaving nitrous acid in solution.

$$Ba(NO_2)_2(aq) + H_2SO_4(aq) \longrightarrow BaSO_4(s) + 2HNO_2(aq)$$

Properties

1 Liquid which is pale-blue because of the presence of N_2O_3.

2 Decomposes rapidly on heating

$$3HNO_2(aq) \longrightarrow HNO_3(aq) + H_2O(l) + 2NO(g)$$
$$2NO(g) + O_2 \text{ from air} \longrightarrow 2NO_2(g)$$

3 Nitrous acid can act as an **oxidizing agent**.

$$2H^+(aq) + 2HNO_2(aq) + 2e^- \longrightarrow 2H_2O(l) + 2NO(g)$$

(i) Iron(II) sulphate in acid solution is oxidized to iron(III) sulphate.

$$2Fe^{2+}(aq) + 4H^+(aq) + 2NO_2^-(aq) \longrightarrow$$
$$2Fe^{3+}(aq) + 2H_2O(l) + 2NO(g)$$
$$2NO(g) + O_2(g) \longrightarrow 2NO_2(g)$$

(ii) Sulphurous acid is oxidized to sulphuric acid.

$$SO_3^{2-}(aq) + 2H^+(aq) + 2NO_2^-(aq) \longrightarrow$$
$$SO_4^{2-}(aq) + 2NO(g) + H_2O(l)$$
$$2NO(g) + O_2(g) \longrightarrow 2NO_2(g)$$

(iii) Iodine is liberated from acidified potassium iodide.

$$2I^-(aq) + 4H^+(aq) + 2NO_2^-(aq) \longrightarrow 2H_2O(l) + 2NO(g) + I_2(s)$$
$$2NO(g) + O_2(g) \longrightarrow 2NO_2(g)$$

4 Nitrous acid can act as a **reducing agent**.

$$HNO_2(aq) + H_2O(l) \longrightarrow HNO_3(aq) + 2H^+(aq) + 2e^-$$

(i) Acidified potassium manganate(VII) is reduced to potassium manganate(II).

$$2MnO_4^-(aq) + 6H^+(aq) + 5NO_2^-(aq) \longrightarrow$$
$$5NO_3^-(aq) + 3H_2O(l) + 2Mn^{2+}(aq)$$

(ii) Acidified potassium dichromate(VI) solution is reduced slowly.

$$Cr_2O_7^{2-}(aq) + 8H^+(aq) + 3NO_2^-(aq) \longrightarrow$$
$$2Cr^{3+}(aq) + 4H_2O(l) + 3NO_3^-(aq)$$

(iii) Bromine water is decolorized.

$$Br_2(aq) + H_2O(l) + NO_2^-(aq) \longrightarrow$$
$$NO_3^-(aq) + 2H^+(aq) + 2Br^-(aq)$$

5   Structure

$$H\text{---}O\text{---}N\text{=}O \quad \text{and} \quad H\text{---}N{\overset{\displaystyle O}{\underset{\displaystyle O}{}}}$$

The free acid is mainly $H\text{---}O\text{---}N\text{=}O$.
As a weak acid, it ionizes only slightly.

$$H\text{---}O\text{---}N\text{=}O(aq) + H_2O(l) \rightleftharpoons$$
$$H_3O^+(aq) + (O\text{=}N \longrightarrow O)^-(aq)$$

Nitric acid, HNO$_3$

Preparation

Distillation of a mixture of potassium nitrate and concentrated sulphuric acid in a retort. The acid distils off and is condensed.

$$KNO_3(s) + H_2SO_4(l) \rightleftharpoons KHSO_4(s) + HNO_3(g)$$

The liquid collected is yellow because of dissolved nitrogen dioxide, which is formed when nitric acid is heated.

$$4HNO_3(l) \longrightarrow 2H_2O(l) + 4NO_2(g) + O_2(g)$$

The acid may be decolorized by bubbling air through it to drive out the nitrogen dioxide.

Production

Ammonia, 10% (from the Haber process), and **air**, 90% by volume, are passed over a **platinum catalyst** at red-heat and the ammonia is oxidized to **nitrogen oxide**.

$$4NH_3(g) + 5O_2(g) \longrightarrow 4NO(g) + 6H_2O(g)$$

The catalyst is heated to start the reaction, which is exothermic. The nitrogen oxide is cooled and passed with excess air through an oxidizing chamber, when nitrogen dioxide is formed.

$$2NO(g) + O_2(g) \rightleftharpoons 2NO_2(g)$$

The temperature has to be kept below 420 K to avoid nitrogen dioxide dissociating. The gas, mixed with excess air is absorbed in water, the temperature being kept high enough (350 K) to avoid the formation of nitrous acid. Nitric acid is produced.

$$4NO_2(g) + O_2(g) + 2H_2O(l) \longrightarrow 4HNO_3(aq)$$

A pressure of about 700 000 Nm^{-2} (7 atm.) is maintained to ensure a flow of gas through the plant.

Physical properties

1 The concentrated acid is a constant boiling mixture containing approximately 35% water.

2 Fuming nitric acid contains 95% acid formed by distilling nitric acid with concentrated sulphuric acid.

3 Structure may be considered to be sp^2 hybridized, the anion is symmetrical and planar.

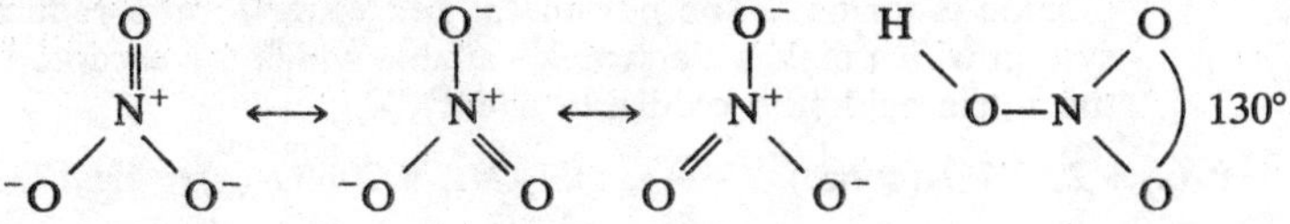

Chemical properties

1 **Acidic properties** The completely anhydrous acid is covalent – no acidic properties. The acidic properties shown in the presence of water are complicated by its oxidizing properties.

(i) Turns blue litmus red.

(ii) Forms salts with bases, e.g.

$$KOH(aq) + HNO_3(aq) \longrightarrow KNO_3(aq) + H_2O(l)$$

(iii) Displaces carbon dioxide from carbonates and hydrogencarbonate, e.g.

$$Na_2CO_3(aq) + 2HNO_3(aq) \longrightarrow 2NaNO_3(aq) + H_2O(l)$$

(iv) Does not liberate hydrogen with any metal – oxidizing properties of nitric acid prevent hydrogen escaping – metallic nitrates, oxides of nitrogen, nitrogen and ammonium compounds formed.

2 **Oxidizing properties of nitric acid**

(i) **With metals** The oxidation state of nitrogen in nitric acid is $+5$. When nitric acid or the nitrate ion accepts electrons, the oxidation state of nitrogen is reduced depending on the products, nitrogen dioxide (nitrogen in $+4$ oxidation state), nitrogen oxide (nitrogen in $+2$ oxidation state) and ammonium ion (nitrogen in -3 oxidation state),

i.e. $NO_3^- \rightarrow NO_2$ diminution in oxidation state of nitrogen form $+5$ to $+4$ corresponding to the acceptance of 1 electron.

$$NO_3^- \longrightarrow NH_4^+ \text{ acceptance of 8 electrons.}$$

The reaction depends on the concentration of the acid and the nature of the metal. Usually several occur at the same time, one predominating, for example: Conc. nitric acid and copper

$$NO_3^-(aq) + e^- + 2H^+(aq) \longrightarrow NO_2(g) + H_2O(l)$$

50% nitric acid and copper

$$NO_3^-(aq) + 3e^- + 4H^+(aq) \longrightarrow NO(g) + 2H_2O(l)$$

Dilute nitric acid and zinc

$$NO_3^-(aq) + 8e^- + 10H^+(aq) \longrightarrow NH_4^+(aq) + 3H_2O(l)$$

(ii) Iron becomes passive and aluminium resists attack possibly due to a protective layer of oxide.

(iii) **With non-metals** Hot concentrated nitric acid oxidizes some non-metals to the most highly oxidized stable acid and nitrogen dioxide is formed. The non-metal and water form a reducing system which makes electrons available which are accepted by the nitric acid (the oxidizing agent), e.g.

$$P_4(s) + 20HNO_3(conc.) \longrightarrow 4H_2O(l) + 20NO_2(g) + 4H_3PO_4(l)$$

$$S(s) + 6HNO_3(conc.) \longrightarrow 2H_2O(l) + 6NO_2(g) + H_2SO_4(l)$$

$$I(s) + 10HNO_3(conc.) \longrightarrow 4H_2O(l) + 10NO_2(g) + 2HIO_3(aq)$$

(iv) Iron(II) is oxidized to iron(III) in iron(II) sulphate in the presence of dilute sulphuric acid and concentrated nitric acid.

$$3Fe^{2+}(aq) + 4H^+(aq) + NO_3^-(aq) \longrightarrow$$
$$3Fe^{3+}(aq) + 2H_2O(l) + NO(g)$$

(v) Hydrogen sulphide is oxidized to sulphur and sulphuric acid.

$$H_2S(g) + 2HNO_3(conc.) \longrightarrow 2H_2O(l) + 2NO_2(g) + S(s)$$

$$H_2S(g) + 8HNO_3(conc.) \longrightarrow 4H_2O(l) + 8NO_2(g) + H_2SO_4(l)$$

(vi) Organic compounds may be oxidized violently, e.g. sawdust or sugar, with nitrogen dioxide evolved. Ethanol becomes ethanedioic acid.

$$C_2H_5OH(l) + 10HNO_3(conc.) \longrightarrow$$
$$(COOH)_2(s) + 7H_2O(l) + 10NO_2(g)$$

(vii) **Alcohols** form esters.

$$ROH(l) + HNO_3(conc.) \rightleftharpoons RNO_3(l) + H_2O(l)$$

(viii) Concentrated nitric and sulphuric acids react with benzene to form nitrobenzene.

Metallic Nitrates

1 All are soluble in water and may be hydrolysed to a basic salt.

2 Decompose on heating.

(i) Ammonium nitrate melts and effervesces forming dinitrogen oxide and steam.

$$NH_4NO_3(s) \longrightarrow N_2O(g) + 2H_2O(l)$$

(ii) Alkali metal nitrates form nitrites and oxygen, e.g.

$$2KNO_3(s) \longrightarrow 2KNO_2(s) + O_2(g)$$

(iii) Other nitrates form a metallic oxide (if stable to heat), e.g.

$$2Pb(NO_3)_2(s) \longrightarrow 2PbO(s) + 4NO_2(g) + O_2(g)$$

(N.B. This is equivalent to the decomposition of nitric acid and $NO_2:O_2$ is always 4:1.)

(vi) When a nitrate is heated and the oxide is unstable, the metal is left, e.g.

$$2AgNO_3(s) \longrightarrow 2Ag(s) + 2NO_2(g) + O_2(g)$$

$$Hg(NO_3)_2(s) \longrightarrow Hg(l) + 2NO_2(g) + O_2(g)$$

Phosphorus, P

Occurrence Found as calcium phosphate(V), $Ca_3(PO_4)_2$, mineral phosphates and in animal bones.

Extraction **Calcium phosphate(V)**, coke and silica (sand) are heated in an electric furnace with **carbon electrodes**. Silicon is a non-volatile acidic oxide and displaces the more volatile phosphorus(V) oxide.

$$2Ca_3(PO_4)_2(s) + 3SiO_2(s) \longrightarrow 6CaSiO_3(l) + P_4O_{10}(g)$$

The **oxide** is reduced by **coke**.

$$P_4O_{10}(g) + 10C(s) \longrightarrow P_4(g) + 10CO(g)$$

Calcium silicate is run off as a molten slag and phosphorus vapour is condensed under water as white phosphorus. The phosphorus is purified by stirring it with warm chromic(VI) acid and then filtering it through canvas. (N.B. The electric current is used to produce heat.)

Allotropy

There are two allotropes, white (or yellow) and red. When white phosphorus is heated in an inert atmosphere (carbon dioxide or nitrogen) at 520 K, it becomes red phosphorus. When red phosphorus is distilled in an inert atmosphere and the vapour is condensed under water, white phosphorus is obtained. White phosphorus is the unstable form, with a higher vapour pressure in all conditions. It is said to be metastable, the white form very slowly changing to the red form.

Monotropy is the direct conversion of one allotrope to another in one direction only. White phosphorus can be changed to red by heating but the reverse change is not direct, requiring a vapour phase inbetween. [*c.f.* the allotropy of sulphur:

$$\text{rhombic sulphur} \underset{\text{below 369 K}}{\overset{\text{above 369 K}}{\rightleftharpoons}} \text{monoclinic sulphur} \Big]$$

In industry, red phosphorus is obtained by heating yellow phosphorus to 520 K in cast-iron vessels with a trace of iodine as a catalyst. Unchanged yellow phosphorus is removed by heating it with sodium hydroxide solution when it forms phosphine, the red allotrope being unaffected. White phosphorus consists of P_4 with phosphorus atoms at the apices of a tetrahedron. The bond angle is $60°$ and this introduces strain into the structure which may account for the reactivity of the allotrope.

Differences between white and red phosphorus

	White phosphorus	Red phosphorus
1	Very slightly soluble in water.	Insoluble in water.
2	Very poisonous.	Not poisonous.
3	Soluble in carbon disulphide.	Insoluble in carbon disulphide.
4	Ignites at 313 K in oxygen.	Ignites at 570 K forming $P_4O_{10}(s)$.

$$P_4(s) + 5O_2(g) \longrightarrow P_4O_{10}(s)$$

5 Ignites spontaneously in chlorine.

$$P_4(s) + 6Cl_2(g) \longrightarrow 4PCl_3(l)$$

With excess chlorine,

$$PCl_3(l) + Cl_2(g) \longrightarrow PCl_5(s)$$

Combines when heated forming PCl_3 and with excess chlorine, PCl_5 is formed.

6 Forms phosphine with hot caustic soda. No reaction.

$$3NaOH(aq) + P_4(s) + 3H_2O(l) \longrightarrow$$
$$PH_3(g) + 3NaH_2PO_2(aq)$$

Other Allotropes **Black phosphorus** is obtained by the action of heat at very high pressure on white phosphorus.
Violet phosphorus is formed when red phosphorus is heated for a long period at 770 K in a vacuum.

Uses In matches, fireworks, smoke-bombs, phosphor-bronze and in rat poison.

Properties of White Phosphorus
Physical properties
Waxy solid, almost colourless when pure but is usually pale yellow, can be cut with a knife, is stored under water, is readily soluble in carbon disulphide and is quite soluble in trichloromethane, turpentine, and ethanol. It is very poisonous, caused 'phossy jaw' in the match industry.

Chemical properties
1 If perfectly dry and pure, white phosphorus can be heated in oxygen without burning, but under ordinary conditions it ignites readily in air. White phosphorus should be handled with care because it may ignite suddenly and cause serious burns.

$$P_4(s) + 5O_2(g) \longrightarrow P_4O_{10}(s)$$

It gives off fumes smelling of garlic and has a greenish glow which can be seen in the dark – the luminescence is due to slow oxidation.

2 With metals Combines with metals in an inert atmosphere to form phosphides, e.g.

$$P_4(s) + 12Na(s) \longrightarrow 4Na_3P(s)$$

3 With non-metals Reaction occurs on heating, e.g.

$$P_4(s) + 5O_2(g) \longrightarrow P_4O_{10}(s)$$

$$P_4(s) + 6Cl_2(g) \longrightarrow 4PCl_3(l)$$

$$P_4(s) + 6S(s) \longrightarrow 2P_2S_3(s)$$

4 White phosphorus is a vigorous **reducing agent**, e.g. precipitates copper from copper(II) sulphate solution.

$$P_4(s) + 10Cu^{2+}(aq) + 12H_2O(l) \longrightarrow$$
$$10Cu(s) + 20H^+(aq) + 4HPO_3(aq)$$

Compounds of phosphorus

Phosphine, PH_3
Preparation White phosphorus is heated with 20% sodium hydroxide solution in an atmosphere of coal gas.

$$P_4(s) + 3NaOH(aq) + 3H_2O(l) \longrightarrow PH_3(g) + 3NaH_2PO_2(aq)$$

The gas is spontaneously inflammable because of the presence of P_2H_4 and the gas is usually not collected but allowed to burn.

$$4PH_3(g) + 8O_2(g) \longrightarrow P_4O_{10}(s) + 6H_2O(l)$$

The gas may be collected over water. It is very poisonous.

Properties Colourless, very poisonous gas, almost insoluble in water, it is not alkaline but forms unstable phosphonium compounds. It is a strong reducing agent and burns readily in air.

Comparison of phosphine and ammonia

	Phosphine	Ammonia
1	Very poisonous.	Kills through asphyxiation.
2	Lower melting point and boiling points because it does not form hydrogen bonds.	Melting and boiling points raised by the formation of hydrogen bonds.
3	Almost insoluble in water.	Very soluble in water.
4	Very much less basic than ammonia – phosphorus atom is bigger than nitrogen atom.	More basic.
5	No alkaline properties.	Alkaline.

$$NH_3(g) + H_2O(l) \rightleftharpoons NH_4^+(aq) + OH^-(aq)$$

6 Phosphonium compounds are unstable, e.g. phosphonium iodide, PH_4I, dissociates on heating and is hydrolysed.

Ammonium salts are more stable, although they decompose on heating, e.g.

$$NH_4Cl(s) \rightleftharpoons NH_3(g) + HCl(g)$$

7 Burns readily in air.

$$4PH_3(g) + 8O_2(g) \longrightarrow P_4O_{10}(s) + 6H_2O(l)$$

Burns when the amount of oxygen present is greater than 25%.

$$4NH_3(g) + 3O_2(g) \longrightarrow 2N_2(g) + 8H_2O(l)$$

8 Strong reducing agent, precipitating copper and silver from their salts, e.g.

$$4Cu^{2+}(aq) + PH_3(g) + 4H_2O(l) \longrightarrow$$
$$4Cu(s) + H_3PO_4(aq) + 8H^+(aq)$$

Shows slight reducing properties, e.g. when passed over strongly heat copper(II) oxide.

$$3CuO(s) + 2NH_3(g) \longrightarrow 3Cu(s) + N_2(g) + 3H_2O(l)$$

Phosphonium iodide, PH_4I. Formed when equimolecular amounts of phosphorus and iodine are dissolved in carbon disulphide. The solvent is distilled off and condensed leaving an intermediate residue of phosphorus and iodine to which water is slowly added. Phosphonium iodide is sublimed off in a stream of carbon dioxide.

$$10I_2(s) + 9P_4(s) + 64H_2O(l) \longrightarrow 20PH_4I(s) + 16H_3PO_4(l)$$

Phosphonium iodide dissociates readily at 303 K,

$$PH_4I(s) \rightleftharpoons PH_3(g) + HI(g)$$

and forms phosphine on heating with alkali, e.g.

$$PH_4I(s) + KOH(aq) \longrightarrow PH_3(g) + H_2O(l) + KI(aq)$$

c.f. $$NH_4Cl(s) + KOH(aq) \longrightarrow NH_3(g) + H_2O(l) + KCl(aq)$$

Oxides and oxy-acids of phosphorus

Oxides

1 Phosphorus(III) oxide, tetraphosphorus hexa-oxide (phosphorus trioxide), P_4O_6.

2 Phosphorus(V) oxide, tetraphosphorus deca-oxide, (phosphorus pentoxide), P_4O_{10}.

Phosphorus(III) oxide, P_4O_6. Obtained when phosphorus is heated in a slow stream of air.

$$P_4(s) + 3O_2(g) \longrightarrow P_4O_6(s)$$

Dissolves in water to form phosphoric acid.

$$P_4O_6(s) + 6H_2O(l) \longrightarrow 4H_3PO_3(s)$$

Phosphorus(V) oxide, P_4O_{10}. Obtained when phosphorus burns in a plentiful supply of air.

$$P_4(s) + 5O_2(g) \longrightarrow P_4O_{10}(s)$$

Properties

1 **With water** Hisses on contact with water forming trioxophosphoric(V) acid.

$$P_4O_{10}(s) + 2H_2O(l) \longrightarrow 4HPO_3(s)$$

This is also formed, slowly, if the oxide is exposed to the air. On boiling with excess water for some time, tetra-oxophosphoric(V) acid is formed.

$$HPO_3(s) + H_2O(l) \longrightarrow H_3PO_4(aq)$$

2 **As a dehydrating agent** Phosphorus(V) oxide is a very powerful dehydrating agent, e.g. when heated with concentrated sulphuric or concentrated nitric acids, sulphur trioxide and dinitrogen pentoxide are formed respectively.

$$2H_2SO_4(l) + P_4O_{10}(s) \longrightarrow 2SO_3(g) + 4HPO_3(s)$$

$$4HNO_3(l) + P_4O_{10}(s) \longrightarrow 2N_2O_5(g) + 4HPO_3(s)$$

Similarly amides are changed to nitrides.

$$2RCONH_2(s) + P_4O_{10}(s) \longrightarrow 2RCN(l) + 4HPO_3(s)$$

Oxy-acids

1 Phosphoric acid (phosphorous acid), H_3PO_3
2 Phosphinic acid (hypophosphorous acid), HPH_2O_2
3 Tetraoxophosphoric(V) acid (orthophosphoric acid), H_3PO_4
4 Heptaoxodiphosphoric(V) acid (pyrophosphoric acid), $H_4P_2O_7$
5 Trioxophosphoric(V) acid (metaphosphoric acid), HPO_3

Phosphoric acid, H_3PO_3. Obtained by the hydrolysis of phosphorus(II) oxide and phosphorus trichloride.

$$P_4O_6(s) + 6H_2O(l) \longrightarrow 4H_3PO_3(s)$$

$$PCl_3(l) + 3H_2O(l) \longrightarrow H_3PO_3(s) + 3HCl(g)$$

The pure acid is a very strong reducing agent, precipitating copper and silver from solutions of their salts, e.g.

$$Cu^{2+}(aq) + H_3PO_3(aq) + H_2O(l) \longrightarrow$$

$$Cu(s) + 2H^+(aq) + H_3PO_4(aq)$$

It is a moderately strong dibasic acid forming acid and normal salts, e.g. sodium hydrogenphosphonate, NaH_2PO_3, and sodium phosphonate, Na_2HPO_3. On heating it disproportionates.

$$4H_3PO_3(s) \longrightarrow 3H_3PO_4(l) + PH_3(g)$$

Phosphinic acid, HPH_2O_2. The sodium salt is obtained when phosphorus is heated with sodium hydroxide solution.

$$P_4(s) + 3OH^-(aq) + 3H_2O(l) \longrightarrow PH_3(g) + 3H_3PO_2{}^-(aq)$$

It is prepared using barium hydroxide and then decomposing the barium salt with dilute sulphuric acid when barium sulphate is precipitated.

$$Ba(H_2PO_2)_2(aq) + H_2SO_4(aq) \longrightarrow BaSO_4(s) + 2HPH_2O_2$$

The pure acid is a crystalline solid. It is a fairly strong monobasic acid. It is a strong reducing agent, e.g.

$$2Cu^{2+}(aq) + 3H_3PO_2(aq) + 3H_2O(l) \longrightarrow$$
$$3H_3PO_3(aq) + 4H^+(aq) + 2CuH(s)$$

Structures

$$\begin{array}{ccc}
\overset{\displaystyle O}{\underset{\displaystyle HO \quad H \quad OH}{\overset{\|}{P}}} & \text{and} & \overset{\displaystyle O}{\underset{\displaystyle HO \quad H \quad H}{\overset{\|}{P}}} \\
H_3PO_3 & & H_3PO_2
\end{array}$$

Only hydroxyl groups can donate a proton.

Tetraoxophosphoric(V) acid, H_3PO_4. Obtained by heating red phosphorus with concentrated nitric acid, which is acting as an oxidizing agent. The reaction is vigorous and phosphorus is added in excess. When cooled, the mixture is treated with distilled water and filtered to remove unused phosphorus. The liquid is evaporated until it is syrupy.

$$P_4(s) + 20HNO_3(conc.) \longrightarrow 4H_3PO_4(l) + 4H_2O(l) + 20NO_2(g)$$

Also formed by the hydrolysis of phosphorus(V) oxide when it is boiled with excess water for some time.

$$P_4O_{10} + 3H_2O(l) \longrightarrow H_3PO_4$$

It is a tribasic acid forming three series of salts, e.g. NaH_2PO_4, Na_2HPO_4 and Na_3PO_4. Ordinary sodium phosphate is $Na_2HPO_4.12H_2O$.

Heptaoxodiphosphoric(V) acid, $H_4P_2O_7$.   This is obtained when tetraoxophosphoric acid is heated.

$$O{=}\overset{\displaystyle OH}{\underset{\displaystyle OH}{P}}{-}\boxed{OH \quad H}O{-}\overset{\displaystyle OH}{\underset{\displaystyle OH}{P}}{=}O \longrightarrow O{=}\overset{\displaystyle OH}{\underset{\displaystyle OH}{P}}{-}O{-}\overset{\displaystyle OH}{\underset{\displaystyle OH}{P}}{=}O$$

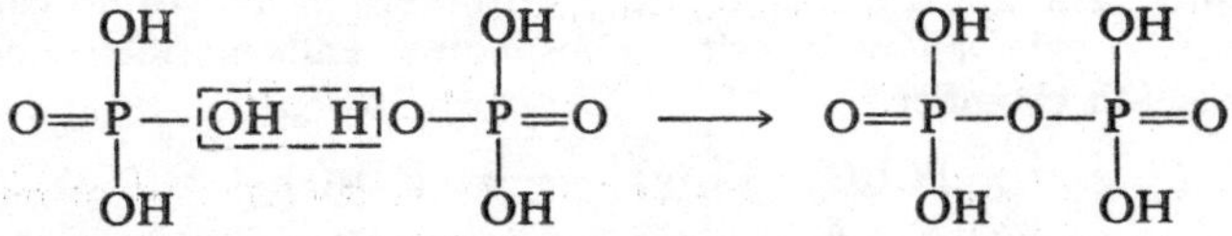

Forms two series of salts, e.g. $Na_2H_2P_2O_7$ and $Na_4P_2O_7$.

Trioxophosphoric(V) acid, HPO_3.   Formed as a polymeric glassy solid when tetraoxophosphoric acid is heated for a long time. It is a monobasic acid.

Relationship between the acids

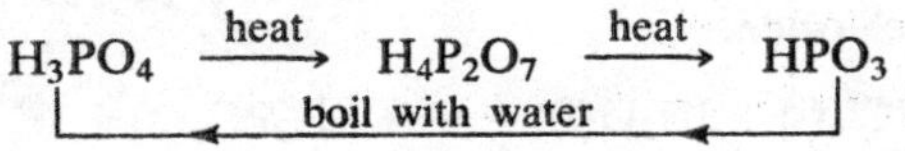

$$H_3PO_4 \xrightarrow{\text{heat}} H_4P_2O_7 \xrightarrow{\text{heat}} HPO_3$$
$$\xleftarrow{\text{boil with water}}$$

Structures

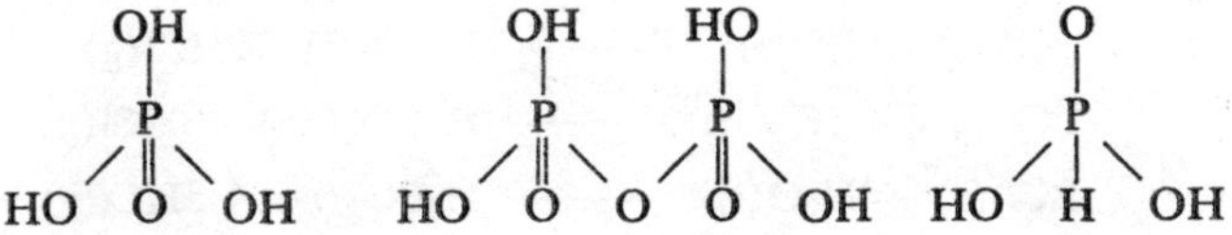

$$H_3PO_4 \qquad H_4P_2O_7 \qquad HPO_3$$

Only hydroxyl groups donate hydrogen ions, H^+.

Chlorides of phosphorus

1 Phosphorus trichloride, PCl_3.
2 Phosphorus pentachloride, PCl_5.

Phosphorus trichloride, PCl_3.   Obtained by the action of chlorine on phosphorus. White phosphorus is heated in a retort by warm water in an atmosphere of chlorine. Air is expelled from the apparatus by carbon dioxide and the phosphorus trichloride distils off and is condensed. The apparatus is kept dry with tubes of anhydrous calcium chloride because any water present would hydrolyse the phosphorus compound.

$$P_4(s) + 3Cl_2(g) \longrightarrow 2PCl_3(s)$$

It is a colourless liquid, fumes in moist air and is hydrolysed by cold water to phosphoric acid.

$$PCl_3(l) + 3H_2O(l) \longrightarrow H_3PO_3(s) + 3HCl(g)$$

Phosphorus pentachloride, PCl_5.   Obtained from phosphorus trichloride, which is allowed to drip slowly into dry chlorine gas in an ice-cooled container. The apparatus is kept dry with tubes of anhydrous calcium chloride. A pale yellow deposit of phosphorus pentachloride collects in the reaction chamber.

$$PCl_3(l) + Cl_2(g) \longrightarrow PCl_5(s)$$

The pentachloride actually consists of PCl_4^- and PCl_6^-. When heated it sublimes and dissociates, the dissociation being complete at 573 K.

$$PCl_5(s) \rightleftharpoons PCl_3(l) + Cl_2(g)$$

Reactions of phosphorus chlorides
Most reactions involve hydroxyl groups.
1   **With water**   Phosphorus trichloride forms phosphoric acid and hydrogen chloride.

$$PCl_3(l) + 3H_2O(l) \longrightarrow H_3PO_3(s) + 3HCl(g)$$

Phosphorus pentachloride forms phosphorus trichloride oxide and then with excess water tetraoxophosphoric(V) acid.

$$PCl_5(s) + H_2O(l) \longrightarrow POCl_3(l) + 2HCl(g)$$

$$POCl_3(l) + 3H_2O(l) \longrightarrow H_3PO_4(aq) + 3HCl(g)$$

or$\qquad PCl_5(s) + 4H_2O(l) \longrightarrow H_3PO_4(aq) + 5HCl(g)$

2   **With alcohols**   Both form corresponding chloroalkanes, e.g. chloroethane.

$$4PCl_3(l) + 6C_2H_5OH(l) \longrightarrow 6C_2H_5Cl(g) + P_4O_6(s) + 6HCl(g)$$

$$PCl_5(s) + C_2H_5OH(l) \longrightarrow C_2H_5Cl(g) + POCl_3(l) + HCl(g)$$

3   **With carboxylic acids**   Both form acyl chlorides, e.g. ethanoylchloride.

$$6CH_3COOH(l) + 4PCl_3(l) \longrightarrow$$
$$6CH_3COCl(l) + P_4O_6(s) + 6HCl(g)$$

$$CH_3COOH(l) + PCl_5(s) \longrightarrow$$
$$CH_2COCl(l) + POCl_3(l) + HCl(g)$$

4   **With carboxylic acids**   Both form acylchlorides, e.g. ethanoylchloride (acetyl chloride).

$$6CH_3COOH(l) + 4PCl_3(l) \longrightarrow$$
$$6CH_3COCl(l) + P_4O_6(s) + 6HCl(g)$$

$$CH_3COOH(l) + PCl_5(s) \longrightarrow CH_3COCl(l) + POCl_3(l) + HCl(g)$$

Structures

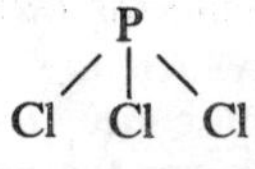

$$PCl_3 \quad \text{Trigonal pyramid} \qquad\qquad PCl_5 \quad \text{Regular trigonal bipyramid}$$

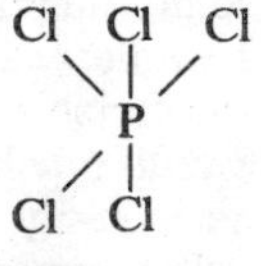

Phosphorus trichloride oxide (phosphoryl chloride or phosphorus oxychloride), $POCl_3$

Formed by distilling phosphorus trichloride with potassium chlorate(V).

$$3PCl_3(l) + KClO_3(s) \longrightarrow 3POCl_3(l) + 3KCl(s)$$

It is manufactured by heating calcium phosphate(V) and charcoal in chlorine at 670 K.

$$Ca_3(PO_4)_2(s) + 6C(s) + 6Cl_2(g) \longrightarrow$$
$$2POCl_3(l) + 3CaCl_2(s) + 6CO(g)$$

which is hydrolysed to tetraoxophosphoric(V) acid.

$$POCl_3(l) + 3H_2O(l) \longrightarrow H_3PO_4(aq) + 3HCl(g)$$

Other phosphorus halides

Trihalides are formed by dissolving phosphorus and bromine or iodine in carbon disulphide and distilling off the solvent.

$$P_4(\text{in } CS_2) + 6Br_2(\text{in } CS_2) \longrightarrow 4PBr_3(\text{in } CS_2)$$
$$P_4(\text{in } CS_2) + 6I_2(\text{in } CS_2) \longrightarrow 4PI_3(\text{in } CS_2)$$

Phosphorus tribromide combines directly with bromine.

$$PBr_3(l) + Br_2(l) \longrightarrow PBr_5(s)$$

No PI_5 exists.

Sulphides of phosphorus Phosphorus combines with sulphur when heated in an inert atmosphere, the quantities as indicated by the formula: P_2S_5, P_4S_3, P_4S_7. Violet phosphorus is dangerously explosive.

Arsenic, As

Occurrence as sulphide As_4S_4 realgar and As_4S_6 orpiment.

Extraction The sulphides are roasted giving the volatile arsenic(III) oxide, As_4O_6, which condenses as a white solid – white arsenic. The oxide is heated with carbon and arsenic sublimes.

$$As_4O_6(s) + C(s) \longrightarrow As_4(s) + 6CO(g)$$

Arsenic is also freed when arsenical pyrites is heated.

$$FeAs(s) + S(s) \longrightarrow FeS(s) + As(s)$$

Properties

Allotropes   Grey arsenic, metallic in appearance, a brittle solid, conducts electricity. When heated, the vapour consists of As_4, above 1073 K, the vapour dissociates to As_2. Rapid cooling of the vapour produces a yellow wax-like solid, **yellow arsenic** which is soluble in carbon disulphide and is easily oxidized with luminiscence (*c.f.* white phosphorus). On gentle warming, yellow arsenic forms the grey allotrope and **black arsenic** is formed as an intermediate. This is also obtained as a mirror-like deposit when arsine is passed through a heated tube. **Grey arsenic** is not oxidized in **air** but burns when heated to arsenic(III) oxide. It is not attacked by dilute acids but dilute nitric acid or concentrated sulphuric nitric acid oxidize it to arsenious acid, H_3AsO_3 and hot concentrated nitric acid oxidizes it to arsenic acid, H_3AsO_4. Unlike phosphorus, arsenic is only slightly attacked by boiling sodium hydroxide solution forming arsenite. No arsine is formed.

$$As_4 + 12OH^- \longrightarrow 4AsO_3^{3-} + 6H_2$$

Used as alloys with heavy metals, e.g. gun-shot is As/Pb.

Compounds of arsenic

Arsine, AsH_3.   Obtained when any arsenic compound is reduced and hydrogen is evolved, e.g. Zn/HCl or Al/NaOH. Prepared by the action of water on sodium arsenide.

$$Na_3As(s) + 3H_2O(l) \longrightarrow 3NaOH(aq) + AsH_3(g)$$

Colourless, very poisonous gas, smelling of garlic. Decomposes on heating to arsenic and hydrogen. The arsenic atom has little power to donate electrons although this property is enhanced by replacement of hydrogen atoms by methyl group; trimethyl arsine, $(CH_3)_3$ As, can form compounds like $(CH_3)_3$ As $\rightarrow BH_3$.

Diarsine, As_2H_4.   Very unstable but $(CH_3)_2As—As(CH_3)_2$, cacodyl is more stable.

Halides of arsenic

Arsenic(III) chloride is obtained when dry hydrogen chloride is passed over arsenic(III) oxide at 473 K.

$$As_4O_6(s) + 12HCl(g) \longrightarrow 4AsCl_3(l) + 6H_2O(l)$$

A volatile liquid not completely hydrolysed by water, arsenious acid being formed.

$$AsCl_3(l) + 3H_2O(l) \rightleftharpoons H_3AsO_3(s) + 3HCl(g)$$

c.f. PCl_3 which is completely hydrolysed. Only penta-halide formed is AsF_5.

Oxides and oxy-acids of arsenic

Arsenic(III), As_4O_6, is an amphoteric solid formed when arsenic burns in air. Exists in two crystalline forms and dissociates to As_2O_3 above 1873 K. Forms arsenious acid with water.

$$As_4O_6(s) + 6H_2O(l) \rightleftharpoons 4H_3AsO_3(l)$$

Very weak acid. **Arsenic(V) oxide**, As_4O_{10} cannot be made directly (*c.f.* P_4O_{10}). Arsenic acid is obtained by oxidizing arsenious acid with nitric acid.

$$2H_3AsO_3(l) + 2HNO_3(aq) \longrightarrow$$
$$2H_3AsO_4(l) + NO(g) + NO_2(g) + H_2O(l)$$

Arsenic(V) oxide is a white deliquescent solid.

Antimony, Sb

Occurs as **stibnite** which is roasted in air to form the oxide which is then reduced by carbon.

$$Sb_2S_3(s) + 5O_2(g) \longrightarrow Sb_2O_4(s) + 3SO_2(g)$$
$$Sb_2O_4(s) + 4C(s) \longrightarrow 2Sb(s) + 4CO(g)$$

Silvery-white metal with probably three allotropes, yellow, black and 'explosive' (the latter form gives out heat when it is scratched and changes to the ordinary form). Antimony is not attacked by hydrochloric or dilute sulphuric acids. Moderately concentrated nitric acid forms antimony(III) oxide, Sb_4O_6, and with concentrated nitric acid antimony(V) oxide, Sb_2O_5(or Sb_4O_{10}), is formed.

Compounds of antimony

Antimony forms **stibine**, SbH_3, and **tri-** and **penta-halides**, $SbCl_3$ and $SbCl_5$. The chlorides are hydrolysed, precipitating basic antimonyl chloride, e.g.

$$SbCl_3(s) + H_2O(l) \rightleftharpoons O{=}SbCl(s) + 2HCl(l)$$

Antimony forms Sb_4O_6 and Sb_2O_5, and also a mixed oxide, Sb_2O_4, containing both trivalent and pentavalent antimony, $Sb^{III}(Sb^VO_4)$, *c.f.* Pb_3O_4. Antimony(III) oxide is amphoteric and antimony(V) oxide is acidic, dissolving in alkali to form $[Sb^V(OH)_6]^-$, *c.f.* aluminates.

Bismuth, Bi

Occurs as the sulphide, Bi_2S_3, and oxide, Bi_2O_3. The metal is reddish-white and relatively inactive. Acids have little effect but nitric acids gives bismuth(III) nitrate, $Bi(NO_3)_3.5H_2O$. When dissolved in water, Bi^{3+} is hydrolysed giving BiO^+ which is usually precipitated with any anion present. Hydrolysis is suppressed by the addition of the appropriate acid, *c.f.* antimony. Bismuth forms bismuthene, BiH_3 (unstable) but $Bi(CH_3)_3$ is more stable. Bismuth forms two oxides, probably Bi_2O_3 and Bi_2O_5. Trioxide is insoluble in alkalis but forms bismuth salts with acids. When a suspension of the hydroxide is oxidized by manganese(VII) the bismuthate ion $[Bi^V(OH)_6]$ is formed. The bismuthate ion is a very strong oxidizing agent. Bismuth forms two halides and BiF_5 is also known.

Practice questions

1 Outline the outer electronic configuration of the elements of group V. Explain why the maximum covalency of nitrogen is lower than for the other elements in the group V.

2 Explain the changes in valencies that occur in group V.

3 What is allotropy? Illustrate your answer with reference to the elements of group V.

4 Outline the number and stability of hydrides, oxides and chlorides in group V.

5 How may nitrogen be prepared in the laboratory? Name any impurities that may be present.

6 How is nitrogen obtained industrially?

7 How does nitrogen react with (a) metals, (b) non-metals, (c) hydrogen chloride?

8 Outline the ways in which nitrogen is (a) removed from the atmosphere, (b) added to the atmosphere?

9 Outline the conditions of the Haber process.

10 How is ammonia prepared in the laboratory?

11 Outline the bonding in (a) NH_3, (b) NH_4^+, (c) NH_4Cl.

12 Describe the role of hydrogen bonding in the properties of ammonia.

13 How does ammonia form complexes?

14 Outline the reaction(s) between ammonia and (a) air, (b) copper(II) oxide, (c) chlorine, (d) sodium, (e) carbon dioxide.

15 Describe the effect of heat on ammonium chloride.

16 Name five oxides of nitrogen, stating the formula of each and whether each is acidic, neutral or basic.

17 Outline the chemical reactions of dinitrogen oxide.

18 Explain why the nitrogen oxide molecule is paramagnetic.

19 Describe the ways in which nitrogen oxide reacts.

20 How may nitrogen dioxide be prepared from lead nitrate?

21 How does nitrogen dioxide react on heating?

22 Outline the reaction between nitrogen dioxide and   (a) water, (b) sodium hydroxide solution,   (c) copper.

23 Name three oxyacids of nitrogen and give the formula of each.

24 How may nitric acid be prepared? Why may it be discoloured?

25 Outline the production of nitric acid from ammonia.

26 State the reactions between nitric acid and   (a) sodium hydroxide,   (b) copper,   (c) sulphur,   (d) iodine.

27 State the reaction(s) between nitric acid, and   (a) iron(II) sulphate,   (b) hydrogen sulphate,   (c) ethanol.

28 Outline the effect of heat on (a) ammonium nitrate,   (b) sodium nitrate,   (c) lead nitrate,   (d) silver nitrate.

29 Describe the allotropy of phosphorus. In what way does it differ from that of sulphur.

30 How does phosphine differ from ammonia?

31 Name two oxides and five oxy-acids of phosphorus.

32 Outline the reactions of the chlorides of phosphorus.

33 Name the hydride of arsenic and antimony, stating the formula of each.

34 Metallic character increases down group V. Explain this with reference to arsenic and antimony.

10 p-BLOCK ELEMENTS OF GROUP VI

Oxygen, sulphur, selenium, tellurium, and polonium have outer electronic configurations of ns^2np^4 and principal oxidation states of $+2$ and $+6$.

O	$1s^2$ $2s^2$	$2p^4$	
S	$1s^2$ $2s^2$	$2p^6$ $3s^2$ $3p^4$	
Se	$1s^2$ $2s^2$	$2p^6$ $3s^2$ $3p^6$ $3d^{10}$ $4s^2$ $4p^4$	
Te	krypton core	$4d^{10}$ $5s^2$ $5p^4$	
Po		$4d^{10}$ $4f^{14}$ $5s^2$ $5p^6$ $5d^{10}$ $6s^2$ $6p^4$	

General properties

1 These elements show electronegative character, the non-metallic properties being strongest in oxygen and sulphur.

2 Polonium shows mainly metallic properties but it is radioactive with a short half-life.

3 All show a covalency of 2 which may be, for example, as $>\!\!O$ or $=\!\!O$; $>\!\!S$ or $=\!\!S$.

4 All show allotropy.

5 Ability to catenate is shown to different degrees. Sulphur may form rings and chains and also polysulphur compounds, e.g. $S_nCl_2(n=3$ to 6) and thionic acids $H_2S_nO_6(n=3$ to 6). Oxygen shows a maximum chain length of 3 in ozone, O_3, and selenium and tellurium form a few short unstable chains, e.g. Se_2Cl_2.

6 The formation of positive ions is prevented by the high ionization energies of these elements (non-metallic character).

7 Oxygen is limited to a valency of 2 because there are no d orbitals available but the other elements can extend their oxidation states to +4 and +6 as in SO_2, SO_3, SF_6.

8 The compound SF_6 actually completes the series SiF_6^{2-}, PF_6^-, SF_6 with fluorine 'exciting' maximum covalency.

9 The tendency for elements to donate electron pairs decreases down the group so that potassium is analogous to bismuth in that the pair of s-electrons behaves like an inert pair.

10 Oxygen forms oxides, O^{2-}, peroxides O_2^{2-}, and superoxides O_2^-, the last two mainly formed with the more electropositive elements of groups I and II.

11 The enthalpy change for O^- to acquire a further electron is high but it is compensated for by the enthalpy change occurring when the solid crystal is produced $(-\Delta H_{Lat})$ and consequently the more electropositive elements form oxides containing O^{2-}.

12 The lattice enthalpies with sulphides, selenides and tellurides make the formation of ionic compounds progressively more difficult.

13 Sulphur forms S^{2-} with the more electropositive elements of groups I and II but most sulphur compounds are covalent with giant molecules with three dimensional layer or chain structures.

14 Both oxygen and sulphur form links of the type $-O-O-$ and $-S-S-$.

15 Oxygen in water and such derivatives as $(C_2H_5)_2O$ can donate one electron pair by forming, for example, $(C_2H_5)_2O \rightarrow BF_3$. Sulphur in hydrogen sulphide and its derivatives shows less tendency to do this.

16 In aqueous solution, the oxide ion, in say Na_2O, becomes the hydroxyl ion OH^-. The sulphide ion as in Na_2S becomes SH^- in aqueous solution. There is usually a further replacement of sulphur by oxygen.

$$S^{2-}(aq) + H_2O(l) \rightleftharpoons HS^-(aq) + OH^-(aq)$$

$$HS^-(aq) + H_2O(l) \rightleftharpoons H_2S(aq) + OH^-(aq)$$

17 **Hydrides** All with the formula H_2A are known and all are poisonous except water. As the electronegativity of group VI elements decreases, bond pairs are further away from the central atom and the repulsive forces between them are progressively weaker and the angle between the hydrogen atoms decreases, e.g. $H_2O(104°30')$ and $H_2Te(89°30')$.

18 Hydrides behave as weak acids and oxides, sulphides, selenides and tellurides can be regarded as salt derivations of these.

19 H_2Se and H_2Te are less stable than H_2S.

20 Many differences in the behaviour of water are due to the strength of the O—H bond, e.g. thermal stability, also hydrogen sulphide is a strong reducing agent, water is not, hydrogen sulphide is a gas and water is a liquid.

21 **Oxides** Group VI elements form several oxides. AO_2 compounds are formed by burning the elements in air. SeO_2 forms infinite covalent chains.

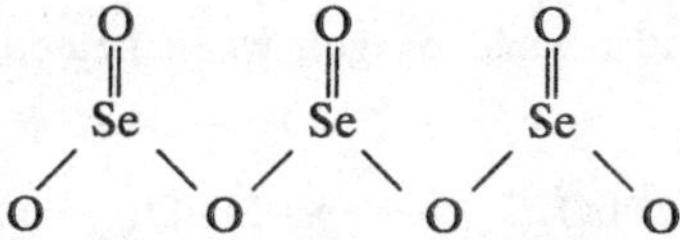

TeO$_2$ has a more ionic structure and PoO$_2$ has two predominantly ionic forms (tetragonal-red) and (face-centred cubic-yellow).

22 Selenic acid, H_2SeO_4, is an oxidizing agent, telluric acid, H_6TeO_6 is also an oxidizing agent, selenious acid, H_2SeO_3 is a poor reducing agent and is not easily changed to selenic acid. This shows that the lower valency states become more stable than in the case of sulphur.

23 **Halides** Those of the type A_2X_2, AX_2 and AX_4 can be formed by direct synthesis and are chemically reactive being hydrolysed by water.

24 The halides, ACl_4, have a distorted trigonal bipyramidal structure with one equatorial position occupied by a lone pair.

25 The ionic character gradually increases from SX_4 to PoX_4. Thus, SCl_4 is a colourless unstable liquid, $SeCl_4$ is a colourless solid, $TeCl_4$ is a white hygroscopic solid with a high melting point and conducts electricity.

Oxygen

Occurrence This is the most abundant element in the Earth's crust. 25% of the atmosphere and 90% of the Earth's water are composed of oxygen. It also occurs as carbonates and silicates in rocks.

Preparation

1 Oxygen is formed when hydrogen peroxide solution is dropped on to manganese(IV) oxide in the cold. The manganese(IV) oxide catalyses the decomposition of the hydrogen peroxide and the gas is collected over water.

$$2H_2O_2(aq) \longrightarrow 2H_2O(l) + O_2(g)$$

2 A mixture of potassium chlorate(V) and manganese(IV) oxide (in the ratio 4:1 by mass) forms potassium chloride and oxygen when heated. The manganese(IV) oxide catalyses the decomposition of the chlorate(V) so that the reaction occurs more quickly and at a lower temperature than when heated alone. The oxygen obtained is often contaminated with chlorine and carbon dioxide, both of which are removed by soda lime, and the oxygen is collected over water.

$$2KClO_3(s) \longrightarrow 2KCl(s) + 3O_2(g)$$

It is likely that potassium manganate(VII) oxide is formed as an intermediate compound.

3 Several oxides also yield oxygen when heated, e.g.

lead(IV) oxide:

$$2PbO_2(s) \longrightarrow 2PbO(s) + O_2(g)$$

mercury(II) oxide:

$$2HgO(s) \longrightarrow 2Hg(l) + O_2(g)$$

4 Electrolysis of acidified water, oxygen being formed at the anode.

Industrially   Oxygen is obtained by the fractional distillation of liquid air. Nitrogen is more volatile and successive distillations yield nearly pure gaseous nitrogen and nearly pure liquid oxygen.

Uses

1 Making steel.
2 Oxyhydrogen and oxyacetylene flames – used for cutting steel-plate and welding.
3 Medicine – with anaesthetics, for carbon monoxide poisoning and in high altitude climbing.
4 In rocket propellent systems.

Physical properties

1 Colourless, odourless gas.
2 Slightly soluble in water – critical for aquatic life.
3 Liquid and solid oxygen are pale blue.
4 Diatomic molecule $O{=}O$.
5 Molecule is paramagnetic, two electrons being unpaired in molecular orbitals.

Chemical properties

1 If the temperature of the gas is raised, the gas dissociates into single atoms and is then very reactive.

2 Forms three ions, O^{2-}, O_2^{2-} and O_2^{-}.

3 **With metals** Oxygen reacts rapidly with electropositive metals to form oxides. Sodium and potassium form monoxide and peroxide. Iron forms black Fe_3O_4. Less electropositive metals react slowly when heated, e.g. lead, copper and mercury. Lead and silver are unchanged.

4 **Non-metals** Many burn readily forming oxides, e.g.

$$S(s) + O_2(g) \longrightarrow SO_2(g)$$

5 **With compounds** Many metallic ores are roasted to form the oxide which is then reduced – this is cheaper than electrical methods, e.g.

$$2ZnS(s) + 3O_2(g) \longrightarrow 2ZnO(s) + SO_2(g)$$

Oxides

1 **Acidic oxides** The oxides of non-metals are usually acidic.
Acid anhydride is an oxide which combines with the elements of water to form an acid, e.g.

$$CO_2(g) + H_2O(l) \rightleftharpoons H_2CO_3(aq)$$

Some react with an alkali to form a salt, e.g.

$$SO_2(g) + 2NaOH(aq) \longrightarrow Na_2SO_3(aq) + H_2O(l)$$

An oxide may be insoluble in water but may still form a salt with an alkali, e.g.

$$SiO_2(s) + 2NaOH(aq) \longrightarrow Na_2SiO_3(aq) + 2H_2O(l)$$

2 **Basic oxides** The oxides of metals are usually basic and react with an acid to form a salt and water only, e.g.

$$CuO(s) + H_2SO_4(aq) \longrightarrow CuSO_4(aq) + H_2O(l)$$

A basic oxide may dissolve in water forming an alkali e.g. Na_2O

3 **Amphoteric oxides** These react with acids and alkalis to form salts, e.g.

$$ZnO(s) + H_2SO_4(aq) \longrightarrow ZnSO_4(aq) + H_2O(l)$$

$$ZnO(s) + 2NaOH(aq) + 3H_2O(l) \longrightarrow Na_2[Zn(OH)_4(H_2O)_2](aq)$$

$$Al_2O_3(s) + 3H_2SO_4(aq) \longrightarrow Al_2(SO_4)_3(aq) + 3H_2O(l)$$

$$Al_2O_3(s) + 2NaOH(aq) + 7H_2O(l) \longrightarrow 2Na[Al(OH)_4(H_2O)_2](aq)$$

4 Neutral oxides   These show neither acidic nor basic properties, e.g. N_2O and CO. (CO can be made to act as the anhydride of methanoic acid.)

5 Peroxides   These are metallic oxides which form hydrogen peroxide with mineral acids. They contain the peroxide link –O–O–, e.g.

$$Na_2O_2(s) + 2HCl(aq) \longrightarrow 2NaCl(aq) + H_2O_2(aq)$$

6 Higher oxides   These contain more oxide than that required by the usual valency of the other elements, e.g. PbO_2, O=Pb=O. They do not contain the peroxide link and may be strong oxidizing compounds.

7 Compound oxides   They behave as if they consist of two different oxides, e.g. Pb_3O_4 behaves as $PbO_2.2PbO$ and black iron oxide Fe_3O_4 as $Fe_2O_3.FeO.$, e.g.

$$Fe_3O_4(s) + 8HCl(aq) \longrightarrow 2FeCl_3(aq) + FeCl_2(aq) + 4H_2O(l)$$

8 Variable valency   This is shown in a number of oxides formed by some elements. In general, the higher oxidation state of a metal forms the more acidic oxide. The acidic character of aluminium oxide is associated with the increase in covalency from 3 in Al_2O_3 to 6 in $Al(OH)_6^{3-}$. Dioxides such as PbO_2 and MnO_2 are strong oxidizing agents because of the higher oxidation state of the metal and not because of the presence of oxygen, (different from peroxides) e.g.

$$PbO_2(s) + 4HCl(conc.) \longrightarrow PbCl_2(s) + 2H_2O(l) + Cl_2(g)$$

Sub-oxides contain the other element in a lower oxidation state than usual, e.g. CO and NO. These are usually reducing agents.

Preparation of oxides

This is dealt with throughout this book. Heating the metal may produce superficial oxidation only, or a number of oxides mixed with nitrides. Usually some other method is used, e.g. the action of heat on carbonate, hydroxide or nitrate.

Effects of heat on oxides

1 Only the oxides of metals at the bottom of the electrochemical series decompose on heating to give the metal, e.g.

$$2HgO(s) \longrightarrow 2Hg(l) + O_2(g)$$

2 An oxide may be changed to another, e.g.

$$2PbO_2(s) \longrightarrow 2PbO(s) + O_2(g)$$

3 Carbon monoxide is more stable than carbon dioxide at the high temperature of the blast furnace and it reduces the ore as it passes through the furnace.

4 The oxides of group I elements are less stable than those of group II elements, otherwise the stability of oxides decreases from left to right across the Periodic Table.

5 Carbon dioxide (consists of discrete molecules) is less stable than silicon dioxide (macromolecular). Solid carbon dioxide has a melting point of 195 K and sublimes, the melting point of silicon dioxide is 1880 K.

Ozone (trioxygen), O_3

Preparation Formed when a silent electrical discharge is passed through dry oxygen. It is found in the upper atmosphere where oxygen is exposed to ultra-violet radiation.

Properties
1 Colourless gas, very poisonous, smells of chlorine and is slightly soluble in water.
2 The molecule is diamagnetic.
3 Endothermic and unstable at room temperature, heating accelerates the decomposition.
4 Ozone is a very powerful oxidizing agent.
   (i) Acidified potassium iodide is oxidized to iodine.

$$2I^-(aq) + 2H^+(aq) + O_3(g) \longrightarrow H_2O(l) + O_2(g) + I_2(s)$$

   (ii) Iron(II) is oxidized to iron(III).

$$2Fe^{2+}(aq) + 2H^+(aq) + O_3(g) \longrightarrow$$
$$2Fe^{3+}(aq) + H_2O(l) + O_2(g)$$

   (iii) Sulphide is oxidized to sulphate.

$$PbS(s) + 4O_3(g) \longrightarrow PbSO_4(s) + 4O_2(g)$$

   (iv) Sulphur dioxide is oxidized to the trioxide.

$$3SO_2(g) + O_3(g) \longrightarrow 3SO_3(g)$$

5 Ozone readily 'adds on' to double bonds in organic compounds, alkenes form ozonides which can be explosive.

$$R_1-CH{=}CH-R_2 + O_3 \longrightarrow R_1-CH\underset{\displaystyle O-\!\!-\!\!-O}{\overset{\displaystyle O}{\diagup \diagdown}}CH-R_2$$

6 Causes mercury to tail, possibly due to oxidation.

Uses Ozone is used to sterilize air in ventilation systems such as in underground railways, in sterilizing water and as a bleach.

Water, H_2O

Physical properties

1   Colourless, odourless, tasteless liquid.
2   Has a maximum density at 277 K. This enables quatic life to survive because ice floats on the surface.
3   Water has high specific heat and latent heats of fusion and vaporization.
4   High dielectric constant – good solvent for ionic compounds.
5   The water molecule is covalent but polar and forms hydrogen bonds with organic compounds containing hydroxyl, carbonyl or amino groups.
6   Ions hydrate easily and the energy of this process enables ionic compounds to dissolve.

Chemical properties

1   Pure water is almost a non-conductor of electricity, slight ionization occurring.

$$2H_2O \rightleftharpoons H_3O^+ + OH^-$$

2   Very stable to heat, decomposed slightly above 2000 K.

3   **With metals**   Electropositive metals react readily with cold water to form hydroxides and hydrogen.

$$Ca(s) + 2H_2O(l) \longrightarrow Ca(OH)_2(aq) + H_2(g)$$

Magnesium reacts with steam.

$$Mg(s) + H_2O(g) \longrightarrow MgO(s) + H_2(g)$$

The reaction with iron at red-heat is reversible, forming iron(II) di-iron(III) oxide.

$$3Fe(s) + 4H_2O(g) \rightleftharpoons Fe_3O_4(s) + 4H_2(g)$$

Aluminium does not react with water unless the protective layer of oxide is broken by, say, mercury.

4   **Non-metals**   Generally no reaction with water although carbon decomposes steam at red-heat.

$$C(s) + H_2O(g) \rightleftharpoons CO(g) + H_2(g)$$

Below 1300 K,

$$C(s) + H_2O(g) \rightleftharpoons CO_2(g) + 2H_2(g)$$

becomes important and similarly silicon at white-heat,

$$Si(s) + 2H_2O \longrightarrow SiO_2(s) + 2H_2(g)$$

Chlorine reacts with water to form chloric(I) acid and hydrochloric acid. Sunlight decomposes the chloric(I) acid.

$$H_2O(l) + Cl_2(g) \rightleftharpoons HCl(aq) + HOCl(aq)$$

$$2HOCl(aq) \longrightarrow 2HCl(aq) + O_2(g)$$

Fluorine attacks water vigorously.

$$2F_2(g) + 2H_2O(l) \longrightarrow 4HF(aq) + O_2(g)$$

$$3F_2(g) + 3H_2O(l) \longrightarrow 6HF(aq) + O_3(g)$$

5 Hydrolysis Water dissociates to form protons and hydroxyl ions. In a solution of salts, the pH depends on the relative strengths of the acids and bases present. Thus the solution of the salt formed from a strong acid and a weak base is acidic, e.g. ammonium chloride.

$$NH_4Cl(aq) + H_2O(l) \rightleftharpoons NH_4OH(aq) + HCl(aq)$$

Hydrochloric acid is a strong acid, almost completely dissociated in water, while ammonium hydroxide is a weak base and only slightly dissociated. Therefore,

$$[H_3O^+] > [OH^-]$$

(i) Chlorides of non-metals are hydrolysed quickly, e.g.

$$PCl_3(l) + 3H_2O(l) \longrightarrow 3HCl(aq) + H_3PO_3(aq)$$

(ii) Chlorides of electropositive metals are not hydrolysed but the chlorides of aluminium and zinc are hydrolysed.

$$ZnCl_2(s) + H_2O(l) \rightleftharpoons Zn(OH)(Cl)aq + HCl(aq)$$

$$AlCl_3(s) + 3H_2O(l) \rightleftharpoons Al(OH)_3(s) + 3HCl(aq)$$

6 Addition reactions Water combines with the oxides of more electropositive oxides to form hydroxides which if soluble are alkalis, e.g.

$$Na_2O(s) + H_2O(l) \longrightarrow 2NaOH(aq)$$

7 Water of crystallization is present in hydrates. In $CuSO_4.5H_2O$, 4 molecules of water are lost on heating and 1 is more strongly held, $[Cu(H_2O)_4]^{2+}SO_4^{2-}.H_2O$.

Hardness of water

Hard water does not lather easily with soap because of the presence of calcium and magnesium ions which form insoluble compounds with soap.

Temporary hardness is caused by the hydrogencarbonates of calcium and magnesium. Rain-water containing dissolved carbon dioxide reacts with limestone rocks forming soluble hydrogencarbonates. This produces caves, pot-holes and gorges.

$$H_2O(l) + CO_2(g) \rightleftharpoons H_2CO_3(aq) \rightleftharpoons 2H^+(aq) + CO_3^{2-}(aq)$$

$$CaCO_3(s) + H_2CO_3(aq) \rightleftharpoons Ca(HCO_3)_2(aq)$$

Permanent hardness is caused by the sulphates of calcium and magnesium.

Softening of water

1   The addition of a calculated quantity of calcium hydroxide (too much causes hardness).

$$Ca(HCO_3)(aq) + Ca(OH)_2(suspension) \longrightarrow$$

$$2CaCO_3(s) + 2H_2O(l)$$

2   The addition of washing soda.

$$Na_2CO_3(s) + CaSO_4(aq) \longrightarrow Na_2SO_4(aq) + CaCO_3(s)$$

3   **Ion exchange**   Natural aluminosilicates, zeolites or synthetic aluminosilicates, e.g. permutit or a cross-linked polymer may be used, e.g. sulphonated polystyrene, with the cation usually Na^+. As the hard water trickles over the ion exchange column, the magnesium and calcium ions change places with the sodium ions leaving the water soft. The material can be regenerated by passing a concentrated solution of sodium chloride through the column, the concentration of sodium ions being high enough to replace the calcium and magnesium held on the column. When the cation on the column is H^+ and the anion is OH^-, **de-ionized water** is formed.

4   Temporary hardness can be removed by boiling when insoluble carbonates are precipitated.

$$Ca(HCO_3)_2(aq) \longrightarrow CaCO_3(s) + CO_2(g) + H_2O(l)$$

Heavy water, D_2O contains deuterium instead of hydrogen.

Hydrogen peroxide, H_2O_2
Preparation   Obtained by the addition of a calculated quantity of **hydrated barium peroxide to ice cold sulphuric acid.**

$$BaO_2.8H_2O(s) + H_2SO_4(aq) \longrightarrow BaSO_4(s) + H_2O_2(aq) + 8H_2O(l)$$

The barium sulphate is filtered off leaving a solution of hydrogen peroxide. The solution can be concentrated up to 60% by heating on a water bath. Further concentration is by distillation with red phosphorus leaving a colourless viscous liquid.

Physical properties
Viscous colourless liquid, although large volumes have a bluish colour.
It is miscible with water in all proportions.

Chemical properties
1 Decomposes when heated and this decomposition is catalysed by
   finely divided substances such as manganese(IV) oxide. It is stored in
   glass bottles and is stabilized by phosphoric acid.

$$H_2O_2(l) \longrightarrow 2H_2O(l) + O_2(g)$$

2 **Powerful oxidizing agent** functioning as an electron acceptor in acidic
   solution.

$$H_2O_2(aq) + 2H^+(aq) + 2e^- \longrightarrow 2H_2O(l)$$

(i) Iron(II) sulphate is oxidized to iron(III) sulphate.

$$2Fe^{2+}(aq) + 2H^+(aq) + H_2O_2(aq) \longrightarrow$$
$$2Fe^{3+}(aq) + 2H_2O(l)$$

(ii) Acidified potassium iodide solution is oxidized to iodine.

$$2I^-(aq) + 2H^+(aq) + H_2O_2(aq) \longrightarrow 2H_2O(l) + I_2(aq)$$

(iii) Sulphurous acid is changed to sulphuric acid.

$$SO_3^{2-}(aq) + H_2O_2(aq) \longrightarrow SO_4^{2-}(aq) + H_2O(l)$$

(iv) Black lead(II) sulphide is oxidized to lead(II) sulphate and this
      is used to restore paintings.

$$S^{2-} + 4H_2O_2(aq) \longrightarrow SO_4^{2-} + 4H_2O(l)$$

(v) Used to bleach hair and as an antiseptic.

3 **As a reducing agent,** hydrogen peroxide acts as an electron donor
   towards several materials which are relatively stronger oxidizing
   agents.

$$H_2O_2(aq) \longrightarrow 2H^+(aq) + O_2(g) + 2e^-$$

(i) Silver oxide forms finely divided silver.

$$Ag_2O(s) + H_2O_2(aq) \longrightarrow 2Ag(s) + H_2O(l) + O_2(g)$$

(ii) Lead(IV) oxide forms lead(II) oxide.

$$PbO_2(s) + H_2O_2(aq) \longrightarrow PbO(s) + H_2O(l) + O_2(g)$$

(iii) Manganese(IV) oxide in dilute sulphuric acid forms a pale pink
      solution of manganese(II) sulphate.

$$MnO_2(s) + H_2O_2(aq) + 2H^+(aq) \longrightarrow$$
$$Mn^{2+}(aq) + 2H_2O(l) + O_2(g)$$

Structure

H–O–OH

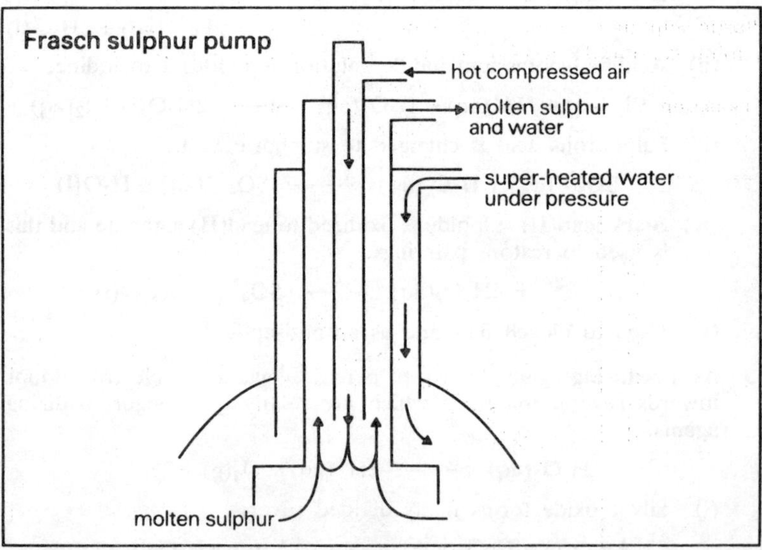

Sulphur, S

Occurrence   Found as free sulphur and as sulphides – PbS (galena), FeS (iron pyrites).

Extraction

Frasch process   The Frasch pump consists of three concentric pipes. Water, superheated under pressure, is forced down the outermost tube and this melts the sulphur. Compressed air is forced down the innermost tube and molten sulphur flows up the middle pipe. It is some 99·8% pure.

Uses   The manufacture of sulphuric acid, the vulcanization of rubber and the manufacture of gunpowder, fireworks and matches.

Allotropy   Solid sulphur has several allotropes, rhombic sulphur (octahedral or α-sulphur), monoclinic sulphur (prismatic or β-sulphur), amorphous sulphur and plastic sulphur. Gaseous sulphur is a mixture of units S_8, S_6, S_4 and S_2 depending on the temperature.

Rhombic sulphur   Powdered sulphur is shaken with carbon disulphide solution and filtered. The filtrate is placed in a beaker covered with filter paper and allowed to evaporate slowly and large crystals of rhombic sulphur are formed.

Monoclinic sulphur   Sulphur is gently heated in an evaporating dish until it is almost full of molten sulphur. It is then allowed to cool until a crust is formed. This is pierced in two places and the liquid poured out. When the crust is removed, the sides of the dish are found to be lined with needle-like crystals of prismatic sulphur.

Amorphous sulphur   Obtained by the acidification of sodium thiosulphate solution or by passing hydrogen sulphide through sulphur dioxide solution.

$$2H_2S(g) + SO_2(aq) \longrightarrow 2H_2O(l) + 3S(s)$$

Plastic sulphur is formed as rubbery coils when a thin stream of molten sulphur is poured into cold water. It is not soluble in carbon disulphide.

Transition $$\text{Rhombic} \underset{\text{below 369 k}}{\overset{\text{above 369 k}}{\rightleftharpoons}} \text{Monoclinic}$$

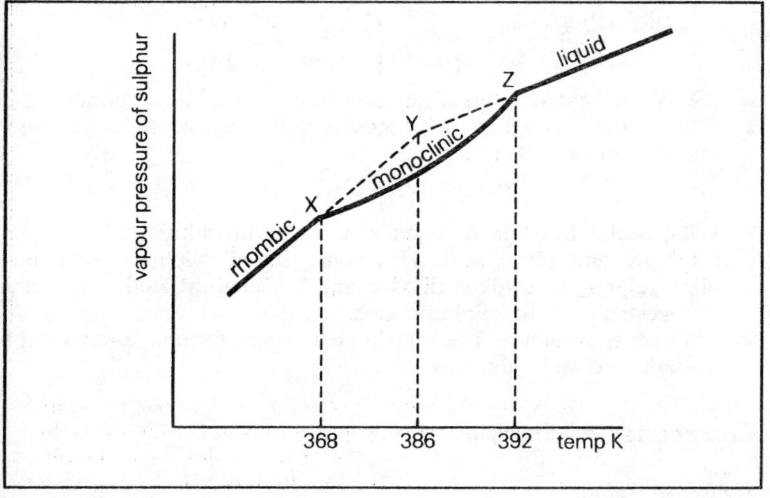

Sulphur shows enantiotropy.
Liquid sulphur   When sulphur is heated out of contact with air:
1   it melts to an amber mobile liquid;

2 on further heating it becomes viscous and will not pour, it darkens;

3 continued heating causes it to become mobile again and to darken until it is nearly black;

4 finally the sulphur boils.

Structure

Rhombic sulphur consists of puckered rings, S_8.

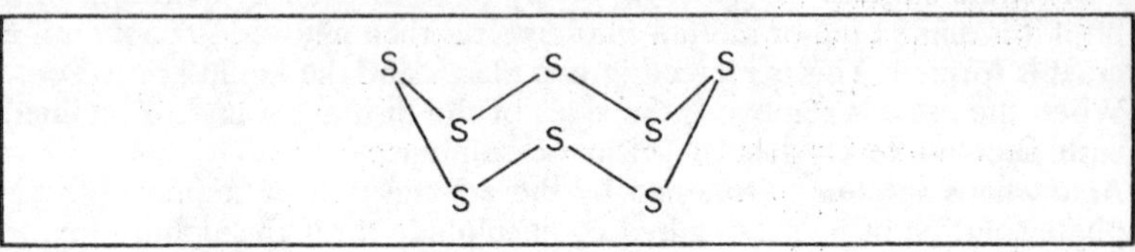

On raising the temperature, the rings break up and the broken chains polymerize. When molten sulphur is cooled rapidly, the S_8 molecules do not have time to form and the open chains entangle forming plastic sulphur.

Chemical properties

1 With non-metals

(i) When heated in oxygen, it burns with a blue flame forming sulphur dioxide.

$$S(s) + O_2(g) \longrightarrow SO_2(g)$$

(ii) When heated in chlorine, disulphur dichloride is formed.

2 With metals Combines with most metals when heated with them out of contact with air, e.g.

$$Fe(s) + S(s) \longrightarrow FeS(s)$$

3 With acids Sulphur is unaffected by hydrochloric acid, dilute sulphuric and nitric acid. Hot concentrated sulphuric acid oxidizes sulphur to sulphur dioxide and hot concentrated nitric acid oxidizes sulphur to sulphuric acid.

4 With aqueous alkali The reaction is complex forming a mixture of thiosulphate and sulphides.

Compounds of Sulphur

Hydrogen Sulphide

Preparation

Obtained by the action of dilute **hydrochloric acid on iron(II) sulphide.**

$$FeS(s) + 2H^+(aq) \longrightarrow H_2S(g) + Fe^{2+}(aq)$$

The gas contains acid and hydrogen because of the presence of free iron as an impurity. The gas may be passed through sodium hydroxide solution which absorbs the acid and hydrogen sulphide allowing the hydrogen to pass on. The sulphide solution is then acidified and the hydrogen sulphide is collected over **hot water**.

Physical properties

Colourless gas, with the smell of rotten eggs, very poisonous.

Chemical properties

1 Burns in a plentiful supply of air with a blue flame.

$$2H_2S(g) + 3O_2(g) \longrightarrow 2H_2O(g) + 2SO_2(g)$$

In a limited supply of air, sulphur is deposited.

$$2H_2S(g) + 2O_2(g) \longrightarrow 2H_2O(g) + 2S(s)$$

2 Hydrogen sulphide is a powerful **reducing agent**, acting as an electron donor, and usually the gas is oxidized to sulphur.

$$H_2S(aq) \rightleftharpoons 2H^+(aq) + S^{2-}(aq)$$

$$S^{2-}(aq) \longrightarrow S(s) + 2e^-$$

(i) **With halogens** When hydrogen sulphide is passed through a solution of a halogen, sulphur is precipitated, e.g.

$$H_2S(g) + Cl_2(aq) \longrightarrow 2HCl(aq) + S(s)$$

(ii) Iron(III) is reduced to iron(II).

$$2Fe^{3+}(aq) + S^{2-}(aq) \longrightarrow 2Fe^{2+}(aq) + S(s)$$

(iii) The purple colour of acidified manganate(VII) solution is discharged and sulphur precipitated.

$$5S^{2-}(aq) + 2MnO_4^-(aq) + 16H^+(aq) \longrightarrow$$
$$2Mn^{2+}(aq) + 8H_2O(l) + 5S(s)$$

(iv) With hydrogen peroxide, sulphur is formed.

$$H_2O_2(aq) + 2H^+(aq) + S^{2-}(aq) \longrightarrow S(s) + 2H_2O(l)$$

(v) Concentrated nitric acid forms sulphuric acid and sulphur.

3 As **an acid** Hydrogen sulphide ionizes slightly.

$$H_2S(aq) \rightleftharpoons H^+(aq) + HS^-(aq) \rightleftharpoons 2H^+(aq) + S^{2-}(aq)$$

It behaves as a weak dibasic acid forming normal and acid salts with caustic alkalis, e.g. Na_2S and $NaHS$. Almost all metallic sulphides are insoluble.

Oxides of sulphur

Sulphur forms a number of oxides, S_2O, S_2O_3, SO_2, SO_3, SO_4.

Sulphur dioxide, SO_2

Preparation   Most metals (not silver or platinum) react with concentrated sulphuric acid to form sulphur dioxide. Usually **concentrated sulphuric** acid is heated with **copper turnings** and the gas is dried by concentrated sulphuric acid and collected by **downward delivery**. Some copper sulphide is also formed.

$$Cu(s) + 2H_2SO_4(l) \longrightarrow CuSO_4(aq) + 2H_2O(l) + SO_2(g)$$

Physical properties
1   Colourless gas with a choking smell.
2   Sulphur dioxide can be liquefied by passing it through a tube cooled by an ice/salt freezing mixture.
3   Very soluble in water.
4   Poisonous.

Chemical properties
1   **As an acidic oxide.**   Sulphur dioxide dissolves in water to form sulphurous acid.

$$SO_2(g) + H_2O(l) \rightleftharpoons H_2SO_3(aq)$$

This is dibasic forming normal and acid salts, e.g. sodium sulphite Na_2SO_3 and sodium hydrogensulphite, $NaHSO_3$.

2   **As a reducing agent.**   Sulphur dioxide acts as an electron donor.

$$H_2O(l) + SO_2(g) \rightleftharpoons 2H^+(aq) + SO_3^{2-}(aq)$$

$$SO_3^{2-}(aq) + H_2O(l) \longrightarrow SO_4^{2-}(aq) + 2H^+(aq) + 2e^-$$

(i)   Iron(III) is reduced to iron(II).
(ii)   Acidified potassium dichromate(VI) is reduced to chromium(III) (orange solution becomes green).
(iii)   Acidified potassium manganate(VII) solution is decolorized.

$$2MnO_4^-(aq) + 6H^+(aq) + 5SO_3^{2-}(aq) \longrightarrow$$
$$2Mn^{2+}(aq) + 3H_2O(l) + 5SO_4^{2-}(aq)$$

(iv)   Chlorine and bromine are reduced to the halogen acid but the reaction with fluorine is reversible.

$$Cl_2(aq) + H_2O(l) + SO_3^{2-}(aq) \longrightarrow$$
$$SO_4^{2-}(aq) + 2Cl^-(aq) + 2H^+(aq)$$

(v)   Iodate(V) is reduced to iodine.

$$2IO_3^-(aq) + 4H^+(aq) + 5SO_3^{2-}(aq) \longrightarrow$$
$$2HSO_4^-(aq) + 3SO_4^{2-}(aq) + H_2O(l) + I_2(s)$$

(vi)   **Bleaching action of sulphur dioxide.**   Coloured dyes are reduced to leuco-compounds (colourless).

$$SO_3^{2-}(aq) + H_2O(l) + X \longrightarrow DO_4^{2-}(aq) + XH_2$$

3   As an oxidizing agent
(i)   Some substances continue to burn in sulphur dioxide e.g.

$$3Mg(s) + SO_2(g) \longrightarrow MgS(s) + 2MgO(s)$$

Potassium, calcium, iron and arsenic will burn if strongly heated forming sulphites and thiosulphates.

(ii) Sulphur dioxide is reduced if passed over heated carbon.

$$C(s) + SO_2(g) \longrightarrow CO_2(g) + S(s)$$

4   Additive reactions
(i)   Lead(IV) oxide glows brightly in sulphur dioxide forming lead sulphate.

$$PbO_2(s) + SO_2(g) \longrightarrow PbSO_4(s)$$

(ii) Chlorine combines directly, the reaction being catalysed by sunlight.

(iii) When heated in the presence of a catalyst such as platinized asbestos, sulphur dioxide combines with oxygen to form sulphur trioxide.

$$2SO_2(g) + O_2(g) \longrightarrow 2SO_3(s)$$

5   As a non-aqueous solvent for iodine, potassium iodide and sulphur. It ionizes positively as

$$2SO_2 \rightleftharpoons SO^{2+} + SO_3^{2-}$$

Lewis acid Lewis base

c.f. $$2H_2O \rightleftharpoons H_3O^+ + OH^-$$

Structure Resonance between

$$O{=}S \overset{}{\underset{\searrow}{}} O \longleftrightarrow S \to O$$

but since the outer quantum shell of sulphur can contain more than 8 electrons, it can be written

$$O{=}S\underset{\searrow}{\diagdown} O$$

Uses
1   Most important is the manufacture of sulphuric acid.
2   Manufacture of sulphites and hydrogensulphites, e.g. calcium hydrogensulphite used to make paper from wood pulp.
3   Bleaching – wool, straw, silk.

Pollution Sulphur dioxide is formed when sulphur compounds are burned. The gas is a pollutant and it also dissolves in rain forming corrosive sulphurous acid.

Sulphur trioxide, SO_3

Preparation Obtained by passing dried **sulphur dioxide** and **oxygen** over a heated **catalyst** of **platinized asbestos**. The gaseous mixture is then passed through a U-tube cooled in a freezing mixture of ice/salt and white needles of sulphur trioxide form in the tube.

$$2SO_2(g) + O_2(g) \longrightarrow 2SO_3(s)$$

Also obtained when some substances are heated, e.g.

$$2FeSO_4(s) \longrightarrow Fe_2O_3(s) + SO_2(g) + SO_3(g)$$

Physical properties

Exists in two forms. **α-sulphur trioxide** is a colourless liquid which can be condensed by ice/water to a solid S_3O_9. When α-sulphur trioxide is kept at room temperature in contact with moist air it forms **β-sulphur trioxide** which is a chain-like polymer.

Chemical properties

1 **With cold water** the reaction is violent and exothermic forming sulphuric acid.

$$H_2O(l) + SO_3(s) \longrightarrow H_2SO_4(l)$$

2 **With metallic oxides.** Combines readily on heating.

$$BaO(s) + SO_3(g) \longrightarrow BaSO_4(s)$$

3 **On heating** it decomposes.

$$2SO_3(g) \rightleftharpoons 2SO_2(g) + O_2(g)$$

Structure Planar with resonance between.

There are twelve electrons in the outer quantum shell of the sulphur atom and the structure may be shown as:

Sulphuric acid (sulphuric(VI) acid), H_2SO_4

Preparation

Lead chamber process   Now obsolete. This was based on the catalytic oxidation of sulphur dioxide in the presence of water with nitrogen oxide as catalyst.

$$NO(g) + \tfrac{1}{2}O_2(g) \longrightarrow NO_2(g)$$

$$NO_2(g) + H_2O(l) + SO_2(g) \longrightarrow H_2SO_4(l) + NO(g)$$

Contact process

Purified sulphur dioxide combines with oxygen of the air to form sulphur trioxide which combines with the elements of water (indirectly) to form sulphuric acid.

$$2SO_2(g) + O_2(g) \longrightarrow 2SO_3(s)$$

Conditions   These are determined by considering **Le Chatelier's Principle**. The formation of sulphur trioxide is favoured by a high pressure, high concentration of oxygen and a low temperature. In practice the **pressure is kept high enough to ensure the flow of reactants** through the plant, the **ratio of $SO_2:O_2$ is kept at 3:2 by volume** – any further increase in the concentration of oxygen would dilute the mixture and slow the reaction down. Lowering the temperature slows down the reaction and the **optimum temperature of 720 K** is maintained with a catalyst. This used to be **platinized asbestos** but has been replaced by cheaper **vanadium(V) oxide** or silica grains covered with complex organometallic material containing vanadium.

Production of sulphur dioxide   Obtained by burning sulphur or sulphides.

Purification of sulphur dioxide   The catalyst is easily poisoned by dust particles or, e.g. arsenic(III) oxide. Large particles settle under gravity, fine particles are removed by electrostatic precipitation and the gases are cooled, washed and dried.

Formation of sulphuric acid   Sulphur trioxide cannot be dissolved directly in water because of the formation of an acid spray. It is passed into **98% sulphuric acid** forming **oleum** or fuming sulphuric acid which is then **diluted**.

$$H_2SO_4(l) + SO_3(g) \longrightarrow H_2S_2O_7(l)$$

$$H_2S_2O_7(l) + H_2O(l) \longrightarrow 2H_2SO_4(aq)$$

Uses   Possibly the most important manufactured chemical.

1   In the manufacture of fertilizers – superphosphate and ammonium sulphate.
2   Cleaning and preparation of metals – pickling.
3   Manufacture of dyes and explosives.
4   Sulphonation of oils – detergents.
5   In lead/acid accumulators.

Physical properties
Colourless, viscous liquid (hydrogen bonding), no smell.

Chemical properties
1   As an acid
   (i)   Completely water-free acid may be covalent with no acidic properties. In the presence of water, it is strongly ionized.

$$H_2SO_4(aq) \rightleftharpoons 2H^+(aq) + SO_4^{2-}(aq)$$

   (ii)   Dilute acid forms hydrogen with electropositive metals.

$$Mg(s) + 2H^+(aq) \longrightarrow Mg^{2+}(aq) + H_2(g)$$

   Mercury, lead, copper and bismuth, and the noble metals, are not attacked by the dilute acid while aluminium, nickel and chromium are usually passive, (layer of metal oxide which if removed then metals show expected reactivity).

   (iii)   Dibasic, forming normal and acid salts, e.g. sodium sulphate Na_2SO_4 and sodium hydrogensulphate, $NaHSO_4$.

   (iv)   Displaces carbon dioxide from carbonates and hydrogencarbonates.

2   As a dehydrating agent
   (i)   Removes the elements of water from, e.g. ethanedioic acid and sugar.

$$(COOH)_2(s) \xrightarrow[\text{conc. } H_2SO_4]{\text{heat}} CO(g) + CO_2(g) + H_2O$$

$$C_6H_{12}O_6(s) \xrightarrow[\text{conc. } H_2SO_4]{\text{heat}} 6C(s) + 6H_2O(g)$$

   (ii)   Removes water of crystallization. Blue copper sulphate crystals become white powdered anhydrous copper sulphate.

3   Sulphuric acid forms **hydrates**, $H_2SO.4H_2O$ and $H_2SO_4.H_2O$.

4   Acid is **dehydrated** by heating with phosphoric(V) oxide.

5   As an oxidizing agent   The hot concentrated acid is a very strong oxidizing agent, e.g.

   (i) $2H_2SO_4(l) + 2e^- \longrightarrow SO_4^{2-}(aq) + 2H_2O(l) + SO_2(g)$

   $Cu(s) + 2H_2SO_4(l) \longrightarrow CuSO_4(aq) + 2H_2O(l) + SO_2(g)$

   Further reduction may occur forming sulphide, free sulphur or hydrogen sulphide.

   (ii)   **With non-metals**   The oxide is formed, e.g.

$$C(s) + 2H_2SO_4(l) \longrightarrow CO_2(g)$$

6   With iron(II) sulphate – hot concentrated acid.

$$2Fe^{2+} + 2H_2SO_4(l) \longrightarrow$$

$$2Fe^{3+}(aq) + 2H_2O(l) + SO_2(g) + SO_4^{2-}(aq)$$

7 **Displaces more volatile hydrochloric and nitric acids** from their salts, e.g.

$$NaCl(s) + H_2SO_4(l) \longrightarrow NaHSO_4(s) + HCl(g)$$

[Cannot be used to produce HBr and HI. There are reducing agents and a number of substances formed, e.g. Br_2, I_2, SO_2, H_2S, S.]

8 **Aromatic hydrocarbons are sulphonated** by the hot (>450 K) concentrated acid, e.g. with benzene.

$$C_6H_6(l) + H_2SO_4(l) \longrightarrow C_6H_5SO_3H(l) + H_2O(l)$$

9 Warm acid forms **esters** with alcohols. (If temperature is too high, dehydration may result.)

$$C_2H_5OH(l) + H_2SO_4(l) \longrightarrow C_2H_5HSO_4(l) + H_2O(l)$$

Structure

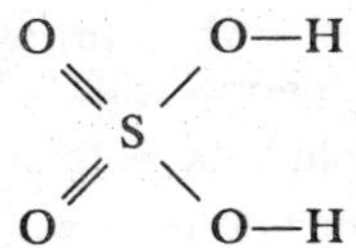

Practice questions

1 What is catenation? Illustrate your answer with reference to group VI.

2 The elements of group VI each show several valencies except oxygen. Discuss this with reference to the electronic configurations of these elements.

3 Write notes on   (a) hydrides,   (b) oxides,   (c) halides, of group VI elements.

4 How may oxygen be obtained from   (a) hydrogen peroxide, (b) potassium chlorate(V)? Name any impurities that may be present and state how they may be removed.

5 What is   (a) an acidic oxide,   (b) an acid anhydride? Give one named example and describe how you would prepare and collect a sample of the oxide.

6 What is   (a) a basic oxide,   (b) an amphoteric oxide? Name one example of each.

7 What is   (a) a peroxide,   (b) a higher oxide,   (c) a mixed oxide? Name one example of each.

8 How does ozone react with   (a) acidified potassium iodide, (b) iron(II) sulphate,   (c) sulphur dioxide?

9 State the reaction(s) between water and   (a) sodium, (b) carbon,   (c) chlorine.

10 Explain why the aqueous solution of ammonium chloride is weakly acid. Name a salt that will dissolve in water to give a basic solution.

11 What is 'hard' water? How can it be softened?

12 What is heavy water?

13 How can a concentrated solution of hydrogen peroxide be prepared?

14 Explain the meaning of '10 vol. hydrogen peroxide'.

15 How does hydrogen peroxide react with   (a) lead(II) sulphide, (b) lead(IV) oxide? What do these two reactions indicate about hydrogen peroxide?

16 How is sulphur extracted by the Frasch pump?

17 Sulphur shows enantiotropy. What does this mean?

18 How does sulphur react with   (a) oxygen,   (b) iron filings, (c) hot concentrated sulphuric acid?

19 How can hydrogen sulphide be prepared?

20 How does hydrogen sulphide react with   (a) chlorine, (b) sodium hydroxide solution?

21 How does sulphur dioxide react with   (a) sodium hydroxide solution,   (b) iron(III) sulphate,   (c) carbon?

22 Outline the production of sulphuric acid.

23 State three ways in which concentrated sulphuric acid may react. Illustrate your answer with one example of each kind of reaction given.

11   p-BLOCK ELEMENTS: HALOGENS OF GROUP VII

Fluorine, chlorine, bromine and iodine have outer electronic configurations of ns^2np^5 and a principal oxidation state of -1. (Astatine is very radioactive.)

F	$1s^2\ 2s^2\ 2p^5$			
Cl	$1s^2\ 2s^2\ 2p^6$	$3s^2\ 3p^5$		
Br	$1s^2\ 2s^2\ 2p^6$	$3s^2\ 3p^6$	$3d^{10}\ 4s^2\ 4p^5$	
I	krypton core	$4d^{10}$		$5s^2\ 5p^5$

(X or X_2 is used to represent a general halogen atom or molecule, Cl, Br, I or Cl_2, Br_2, I_2.)

General properties

1   Very reactive, electronegative elements.

2   All form coloured, diatomic molecules.

3   Melting points and boiling points rise on descending the groups so that, fluorine and chlorine are gases, bromine is a liquid and iodine is a solid.

4   Astatine, the last member of the group, is a volatile solid and radioactive with a short half-life.

5   Fluorine is the most electronegative element in the Periodic Table and shows the oxidation state -1 only but others show several positive oxidation states, e.g. $+7$ for iodine in IF_7 and for chlorine in Cl_2O_7. The higher oxidation states involve the unpairing of p-electrons and their promotion to d-level.

6   The halogens show little tendency to form positive ions.

7   Because of the strong attractive force between the nuclear protons and outer electrons, the halogen atoms are the smallest ones in their respective periods.

8   The addition of an electron produces an increase in size so that the halogen ion X^- is nearly twice the size of the respective atom, X.

9   All halogens show a covalency of 1 with non-metals.

10   The trihalide ions, Br_3^- and Cl_3^- are unstable but I_3^- forms stable ionic salts such as KI_3.

11   Molten iodine conducts electricity, possibly due to self ionization into I_3^+ and I_3^-.

12   All form compounds with each other and polyhalide ions, e.g. in $KICl_2$.

13   Electronegativity decreases down the group.

14   Fluorine is more reactive than the other halogens.
   (i)   The bond enthalpy of F—F is very low because of the repulsive forces between the non-bonding electrons in the molecules.
   (ii) The atom is very small.

15   The small covalent radius of fluorine and its low bond enthalpy means it is able to extend the covalency of other elements to a maximum as in SF_6 and IF_7.

16   The halogens combine directly with most elements.

17   Halogens dissolve in alkalis to form oxy-acid anions (not fluorine).

18   Powerful oxidizing agents because of their ability to acquire electrons – oxidizing power decreases down the group.

19   The energy change for $X(g) + e^- \rightarrow X^-$ is a maximum for chlorine but fluorine is the strongest oxidizing agent because of its low enthalpy and high enthalpy of hydration.

20   Fluorine displaces all the other halogens from their salts, chlorine displaces bromine and iodine and bromine displaces iodine.

21   The halides formed with hydrogen, HX, are colourless, covalent gases, except HF which is a liquid because of the effect of strong hydrogen bonding.

22 Halogens react with hydrogen in the presence of sunlight – occurs via free radical and chain mechanism.

23 Affinity for hydrogen and stability of HX decreases down the group.

24 Aqueous solutions of HX are strong monobasic acids but HF is only partially dissociated and shows weak acidic properties. In concentrated solutions HF forms $[HF_2]^-$.

25 The aqueous solution of HI behaves as a strong reducing agent.

26 Halogens form several oxides which are generally unstable.

27 **Monoxides** OF_2 is a colourless gas, soluble in water forming a neutral solution. Cl_2O is a yellow gas, Br_2O is a brown liquid – acidic oxides.

28 **Dioxides** BrO_2 and ClO_2 are strong oxidizing agents.

29 I_2O_5 is the most stable oxide of the halogens, it is the anhydride of iodic(V) acid.

30 All halogens form oxo-acids, HOX, except fluorine, which are unstable and unknown in the pure state.

31 Acid, HXO_2, is only known for chlorine, $HClO_2$ and is unstable.

32 HXO_3 acids are stable, stability decreasing down the group.

33 HXO_4 known only for chlorine, $HClO_4$.

34 Iodine differs from the other halogens.

35 The iodine atom is larger than the other halogens so that there is not always room for as many iodine atoms as other halogens, e.g. PI_5 does not exist.

36 Intermolecular forces increase with size – iodine is a solid and is soluble in organic solvents.

37 Because the atom is so large, iodine can lose an electron to form I^+ as in ICNO.

38 Silver halides are insoluble, except the fluoride. AgCl is white and soluble in ammonia. AgBr is pale yellow and slightly soluble in ammonia, AgI is yellow and insoluble in ammonia.

39 Bleaching action of the halogens decreases down the group and iodine has no bleaching action.

Chlorine (bromine and iodine)

Occurrence Main source is salts, e.g. sodium and potassium chloride, magnesium chlorides and bromides in sea water but iodine is found combined in seaweed and in saltpetre (as $NaIO_3$).

Preparation

1 **Halide** (NaCl, NaBr or NaI) is heated with manganese(IV) oxide and **concentrated sulphuric acid**. Chlorine gas is collected by downward delivery, bromine liquid is condensed in a cooled flask, iodine is collected on a cooled surface and purified by sublimation.

$$2X^-(s) + 4H^+ + MnO_2(s) \longrightarrow Mn^{2+}(aq) + 2H_2O(l) + X_2$$

2 Chlorine is usually made by the oxidation of **concentrated hydrochloric acid by potassium manganate(VII) in the cold**. The chlorine is washed with water to remove any volatile hydrogen chloride, dried by concentrated sulphuric acid and collected by **downward delivery**.

$$2KMnO_4(s) + 16HCl(aq) \longrightarrow$$
$$2KCl(aq) + 2MnCl_2(aq) + 8H_2O(l) + 5Cl_2(g)$$

Industrial preparation

1 **Chlorine** is obtained by the electrolysis of concentrated sodium chloride in the Kellner-Solvay cell and as the by-product of several industrial processes.

2 Bromine is displaced from sea water by passing chlorine through it.

3 Sodium iodate(V) is reduced by sulphur dioxide. The iodide formed reacts with further iodate.

$$IO_3^-(aq) + 3SO_3^{2-}(aq) \longrightarrow I^-(aq) + 3SO_4^{2-}(aq)$$
$$IO_3^-(aq) + 5I^-(aq) + 6H^+(aq) \longrightarrow 3H_2O(l) + 3I_2(s)$$

Uses

1 **Chlorine** is used as a bleaching agent, in the manufacture of sodium chlorate(I), sodium chlorate(V), hydrogen chloride, organic compounds and for sterilizing water.

2 **Bromine** is used to manufacture dyes and drugs in photography and in 1:2 dibromomethane $BrCH_2.CH_2Br$ which is added to petrol.

3 **Iodine** is used as an antiseptic in solution (tincture of iodine) and as iodoform.

Physical properties

1 Chlorine – yellow-green gas, bromine – red-brown liquid, iodine – dark violet solid which sublimes.

2 Slightly soluble in water, iodine soluble in covalent solvents.

3 All form diatomic molecules.

Chemical properties

1 **Very reactive, electronegative elements**, chlorine is the most reactive displacing bromine and iodine. Iodine is the least reactive.

2 **Univalent**, forming covalent and ionic compounds.

3 **As oxidizing agents** $X_2 + 2e^- \rightarrow 2X^-$
(i) Iron(II) to iron(III), e.g.

$$2Fe^{2+}(aq) + Cl_2(g) \longrightarrow 2Fe^{3+}(aq) + 2Cl^-(g)$$

Iodine does not oxidize Fe^{2+} and Fe^{3+} oxidizes potassium iodide to iodine.

(ii) When **hydrogen sulphide** is passed through water containing a halogen, the colour is discharged and sulphur is precipitated, e.g.

$$H_2S(g) + Cl_2(aq) \longrightarrow 2HCl(aq) + S(s)$$

(iii) **Sulphur dioxide** (sulphurous acid or sulphite) is oxidized to sulphate or sulphuric acid, e.g.

$$SO_2(g) + H_2O(l) \rightleftharpoons 2H^+(aq) + SO_3^{2-}(aq)$$

$$SO_3^{2-}(aq) + Cl_2(aq) + H_2O(l) \longrightarrow$$
$$SO_4^{2-}(aq) + 2H^+(aq) + 2Cl^-(aq)$$

(iv) **Iodine** dissolved in potassium iodide solution oxidizes sodium thiosulphate to sodium tetrathionate.

$$2S_2O_3^{2-}(aq) + I_2(\text{in KI solution}) \longrightarrow S_4O_6^{2-}(aq) + 2I^-(aq)$$

4   Bleaching action of chlorine occurs only in the presence of water. Dry chlorine does not bleach.

$$Cl_2(aq) + H_2O(l) \rightleftharpoons HCl(aq) + HOCl(aq)$$

$$HOCl(aq) + \text{dye} \longrightarrow HCl(aq) + \text{oxidized dye}$$

5   With alkalis

(i)   With **cold dilute alkali**, chlorine and bromine form chloride and chlorate(I) or bromide and bromate(I) respectively.

$$2OH^-(aq) + X_2(g) \longrightarrow X^-(aq) + OX^-(aq) + H_2O(l)$$

Sodium chlorate(I), NaOCl, is a strong oxidizing agent used as a bleach and antiseptic.

(ii) **With hot concentrated alkali**, chloride and chlorate(V) or bromide and bromate(V) or iodide and iodate(V) are formed.

$$6OH^-(aq) + 3X_2(g) \longrightarrow 5X^-(aq) + XO_3^-(aq) + 3H_2O(l)$$

(iii) **Bleaching powder** is formed by passing chlorine over slaked lime. Exact formula is unknown and in the presence of dilute hydrochloric or sulphuric acids, bleaching powder behaves as a chlorinating or oxidizing agent.

$$Ca(OH)_2(s) + Cl_2(g) \longrightarrow CaOCl_2 \cdot H_2O(s)$$

$$CaOCl_2(s) + 2HCl(aq) \longrightarrow CaCl_2(aq) + H_2O(l) + Cl_2(g)$$

6   With hydrogen   Chlorine combines explosively with hydrogen in sunlight, bromine reacts slowly and the reaction between iodine and hydrogen is reversible.

$$H_2(g) + Cl_2(g) \longrightarrow 2HCl(g)$$

7   **With turpentine**   When filter paper, saturated with warm turpentine is dropped into chlorine, hydrogen chloride and black particles of carbon are formed.

$$C_{10}H_{16}(l) + 8Cl_2(g) \longrightarrow 10C(s) + 16HCl(g)$$

8   When **aqueous ammonia** is dropped into chlorine gas, the mixture reacts vigorously giving out green sparks.

$$8NH_3(\text{conc. aq}) + 3Cl_2(g) \longrightarrow 6NH_4Cl(s) + N_2(g)$$

With **excess chlorine** the dangerously explosive oil NCl_3 may be formed. Bromine behaves similarly, iodine forms a brown compound $NI_3.nNH_3$.

9   When **sulphur** is heated in dry chlorine, disulphur dichloride is formed. The reaction with bromine is similar but sulphur does not combine with iodine.

$$2S(s) + Cl_2(g) \longrightarrow S_2Cl_2(l)$$

10   **White phosphorus** burns spontaneously in chlorine and **red phosphorus** on heating forms trichloride or in presence of excess chlorine, the pentachloride is formed.

$$P_4(s) + 6Cl_2(g) \longrightarrow 4PCl_3(l)$$

Liquid bromine inflames when dropped on to red phosphorus.

11   The halogens do not combine directly with **oxygen, nitrogen,** or **carbon.**

12   The halogens combine directly with most metals, especially if heated, the metal forming its ion with the highest oxidation state, e.g.

$$2Fe(s) + 3Cl_2(g) \longrightarrow 2FeCl_3(s)$$

Mercury and iodine react if rubbed together in a pestle and mortar forming red mercury(II) iodide, HgI_2 or green mercury(I) iodide, Hg_2I_2.

13   **Anhydrous chlorides** cannot be crystallized because they are usually hydrolysed.
(i)   **Dry chlorine or hydrogen chloride** are passed over the heated metal, the apparatus being sealed with tubes of anhydrous calcium chloride.
(ii)   **Insoluble chlorides** are precipitated by hydrochloric acid.
(iii) **Neutralization**   Hydroxide may be titrated against dilute hydrochloric acid using litmus as indicator, the litmus being removed by carbon and filtered off. Filtrate evaporated to dryness – some hydrolysis may occur.

$$NaOH(aq) + HCl(aq) \longrightarrow NaCl(aq) + H_2O(l)$$

(iv) Dilute hydrochloric acid may be added to carbonates, e.g.

$$CaCO_3(s) + 2HCl(aq) \longrightarrow CaCl_2(aq) + CO_2(g) + H_2O(l)$$

Hydrogen chloride is prepared by the action of **cold concentrated sulphuric acid on sodium chloride** (or rock salt). The hydrogen chloride is dried with calcium chloride and collected by downward delivery.

$$NaCl(s) + H_2SO_4(l) \longrightarrow NaHSO_4(s) + HCl(g)$$

It is **manufactured** by the direct combination of **hydrogen** and **chlorine** in the presence of **activated charcoal**.

$$H_2(g) + Cl_2(g) \longrightarrow 2HCl(g)$$

When **dry, hydrogen chloride is covalent and not acidic**. In the presence of water,

$$HCl(aq) \rightleftharpoons H^+(aq) + Cl^-(aq)$$

In solution, the acid turns blue litmus red, liberates hydrogen from metal, forms salts with bases and displaces carbon dioxide from carbonates. Concentrated hydrochloric acid is oxidized by strong oxidizing agents to chlorine.

Bonding in chlorides

Electropositive metals form chlorides with ionic lattices – non-volatile solids, e.g. sodium chloride, barium chloride.

Non-metals, e.g. boron, carbon, silicon and sulphur, form covalent chlorides – volatile liquids.

Some elements form both ionic, and for the higher valency, covalent chlorides, e.g. $PbCl_2$(ionic) and $PbCl_4$(covalent).

where $M = Fe^{3+}$, Al^{3+}, Au^{3+}

Oxides and oxy-acids of the halogens   Chlorine forms six oxides – all unstable, e.g. Cl_2O is a brown gas which explodes on warming; ClO_2 is an orange-yellow gas; Cl_2O_6 is a liquid and is not paramagnetic(ClO_3 would be paramagnetic); Cl_2O_7 is a colourless oily liquid. Oxyacids include $HOCl$, $HClO_2$, $HClO_3$ and $HClO_4$.

Practice questions

1 The halogens form ionic and covalent bonds. Explain why this is possible, illustrating your answer with reference to the molecule HCl, when dry and in the presence of water.

2 Write notes on the oxides and oxy-acids formed by the halogens.

3 How may chlorine be prepared and collected in the laboratory?

4 Which of the following reactions are not possible?

(a) $\qquad$ $2NaCl + Br_2 \longrightarrow 2NaBr + Cl_2$

(b) $\qquad$ $2NaI + Cl_2 \longrightarrow 2NaCl + I_2$

(c) $\qquad$ $2NaBr + I_2 \longrightarrow 2NaI + Br_2$

(d) $\qquad$ $2NaI + Br_2 \longrightarrow 2NaBr + I_2$

5 Give the reaction(s) between chlorine and (a) hydrogen sulphide, (b) cold, dilute sodium hydroxide solution, (c) hot concentrated sodium hydroxide solution.

6 Give the reaction(s) between chlorine and (a) ammonia, (b) sulphur, (c) phosphorus.

7 How may hydrogen chloride be prepared and collected in the laboratory?

12 d-BLOCK ELEMENTS: TRANSITION ELEMENTS

These consists of three series, each of ten elements and a fourth incomplete series. These notes refer to the first series scandium, titanium, vanadium, chromium, manganese, iron, cobalt, nickel, copper and zinc.

Sc		$3d^1 4s^2$	Fe		$3d^6\ 4s^2$
Ti		$3d^2 4s^2$	Co		$3d^7\ 4s^2$
V	argon core	$3d^3 4s^2$	Ni	argon core	$3d^8\ 4s^2$
Cr		$3d^5 4s^1$	Cu		$3d^{10} 4s^1$
Mn		$3d^5 4s^2$	Zn		$3d^{10} 4s^2$

General properties

1 The transition elements are formed by the filling of the penultimate d-orbitals.
2 They are all metals.
3 They have similar atomic radii.
4 Because they are so close in size, they form alloys with each other and with other metals.
5 They form interstitial compounds which are non-stoichiometric, e.g. steel contains interstitial carbon.
6 The outer electrons are pulled in by the increased nuclear charge and their densities are greater than for s-block elements.
7 The melting and boiling points are generally higher than those of

the s-block elements. When the d-sublevel is completely filled the boiling points are lower as with copper and zinc.

8 The increase in the nuclear charge holds the outer electrons of the d-block elements more firmly than those of the s-block. They have higher ionization energies but the ionization energies are still small.

9 Electronegativity increases across the Periodic Table but the last member of the series has a lower value than the element preceding it.

10 The tendency to form covalent bonds increases as the atoms get smaller, from Sc → Cu.

11 They show variable valency because of the ease with which the d-electrons can be promoted.

12 In chromium and copper, electronic stability is acquired by the movement of an s-electron to the d-level giving chromium a half filled d-level (5 electrons) and copper a filled d-level (10 electrons).

13 These elements may show an apparent zero valency as in the carbonyls, e.g. $Ni(CO)_4$.

14 The stability of the highest oxidation states shown by the elements decreases from Sc → Zn and those compounds containing the higher oxidation states tend to be oxidizing agents.

15 The oxides formed by these metals in their higher oxidation states tend to be acidic and covalent.

16 The oxides formed by these metals in their lower oxidation states tend to be basic and ionic.

17 Most transition metals have unpaired electrons – paramagnetic. Iron, cobalt and nickel show ferro-magnetism.

18 Form coloured ions. This arises from the promotion of d-electrons. The electronic transitions which are possible correspond to energy of a definite wavelength and this determines the light which is absorbed or reflected. (Most s-block metals or metal ions are white or colourless).

19 Show non-stoichiometry, especially in sulphides and oxides. Zinc oxide becomes yellow when heated because of lattice defects.

20 They show catalytic properties, e.g. Pt, Ni, Fe and Pd. They may provide a suitable surface or may form unstable intermediates – facilitated by their variable oxidation states.

21 They form many complexes – see section on complexes.

22 Oxidation states:

Sc	Ti	V	Cr	Mn	Fe	Co	Ni	Cu	Zn
				+7					
			+6	+6	+6				
		+5	+5	+5	+5				
	+4	+4	+4	+4	+4	+4	+4		
+3	+3	+3	+3	+3	+3	+3	+3	+3	
	+2	+2	+2	+2	+2	+2	+2	+2	+2
			+1					+1	

Iron

Occurrence   Found as **haematite**, Fe_2O_3, **magnetite**, Fe_3O_4 and carbonate, $FeCO_3$.

Extraction

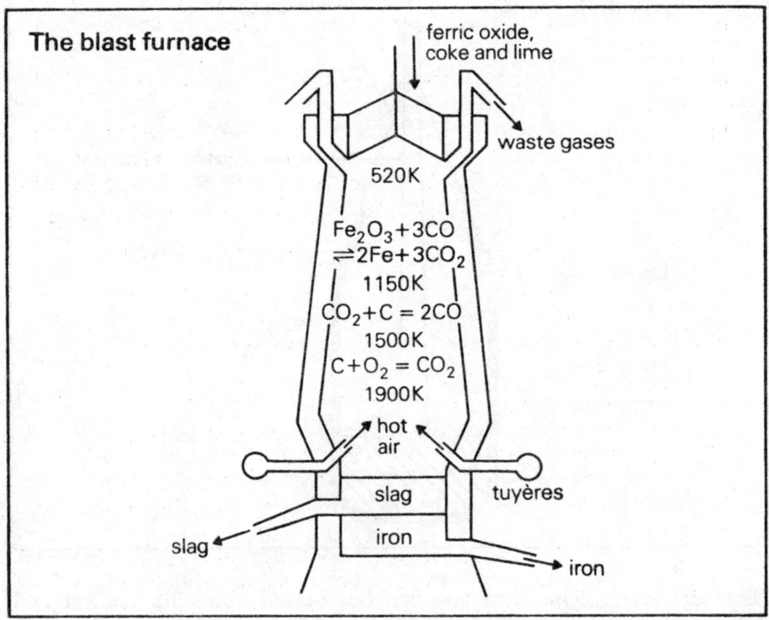

A mixture of **coke, limestone** and **iron ore** is heated in a blast furnace by hot oxygen-enriched air. Near the base of the furnace, the **coke burns to carbon monoxide.**

$$C(s) + O_2(g) \longrightarrow CO_2(g)$$

$$CO_2(g) + C(s) \longrightarrow 2CO(g)$$

The reaction is exothermic and carbon dioxide is not reduced below 973 K. Near the top of the furnace, the hot **carbon monoxide reduces the iron oxide.**

$$Fe_2O_3(s) + 3CO(g) \longrightarrow 2Fe(l) + 3CO_2(g)$$

The limestone decomposes to calcium oxide and carbon dioxide.

$$CaCO_3(s) \rightleftharpoons CaO(s) + CO_2(g)$$

The quicklime combines with earthy materials to form a slag that floats on top of the molten iron collecting at the bottom of the furnace. The molten metal is run into moulds or pigs and solidifies as pig-iron.

Basic oxygen steel making (B.O.S.)

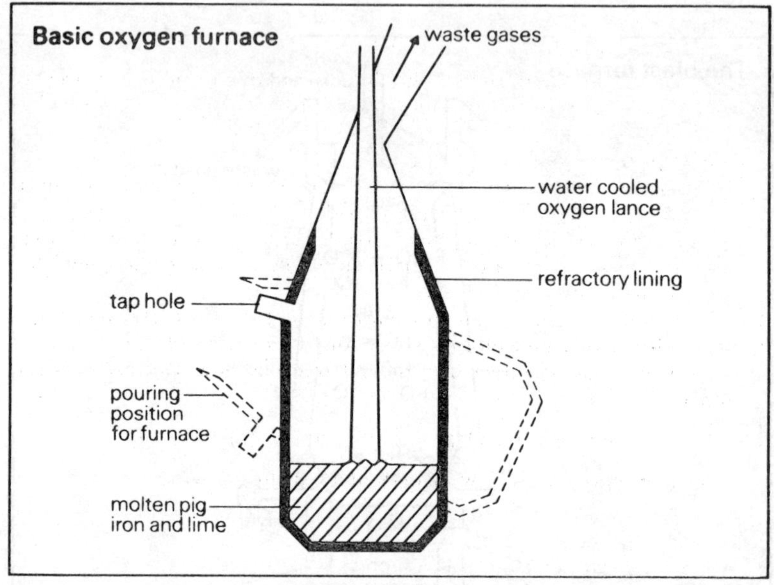

The furnace is cylindrical and can be rotated about its vertical and horizontal axes. The tilted furnace is packed with pig-iron and 30% scrap iron and then rotated into its vertical position. It is heated and a blast of oxygen under great pressure is directed on to the surface of the molten charge by a water-cooled lance. To prevent the lance burning away, the oxygen is diluted with air, steam or carbon dioxide. The blow lasts for some 15 minutes and during it, lime is added as a flux to remove oxidized impurities as a slag, e.g. oxides of silicon, phosphorus, and manganese. When the composition of the steel is correct, the blow is turned off, the furnace tilted and the metal run off from a tap-hole. The remaining slag is used as part of the next charge.

Cast iron   Impure iron containing C, P, Si, Mn and S. It is brittle and used to cast small objects, e.g. gas stoves, gas rings and manhole covers.

Wrought iron   Pure iron obtained by heating cast iron on a hearth lined with haematite – very mild steel.

Steel This is iron containing calculated qualities of other elements such as carbon and manganese.
1.5% C – high tensile steel; 85% iron, 15% Cr – stainless steel.

Tempering or annealing of steel is the heating and controlled quenching of the metal so as to vary its hardness.

Properties of Iron
Physical properties
Grey metal which is soft, malleable, ductile, highly magnetic – ceases to be magnetic at 1042 K. Three allotropes exist.

Chemical properties
1 It is not affected by dry air or dry oxygen at room temperature but if heated forms iron(II) diiron(III) oxide, Fe_3O_4.
2 When iron filings and sulphur, mixed in the ratio of their relative atomic masses, are heated, iron(II) sulphide, FeS, is formed with the evolution of heat.
3 The metal combines with dry chlorine to form iron(III) chloride.
4 With dry hydrogen chloride, iron forms iron(II) chloride.
5 Pure water has no effect on iron. At red-heat steam forms iron(II) diiron(III) oxide.

$$3Fe(s) + 4H_2O(g) \rightleftharpoons Fe_3O_4(s) + 4H_2(g)$$

6 Iron combines directly with carbon to form a carbide, Fe_3C.

Compounds of iron

Iron has a valency of 2 or 3.

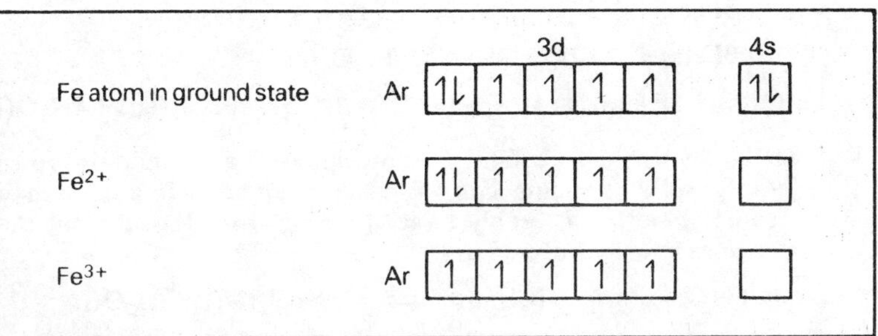

The comparative stability of Fe(III) is associated with its electronic structure in which the 3d-subshell is half-filled.

Iron(II) compounds are formed when the metal is heated with acids (not nitric acid), e.g.

$$Fe(s) + H_2SO_4(aq) \longrightarrow FeSO_4(aq) + H_2(g)$$

$$Fe(s) + 2HCl(g) \longrightarrow FeCl_2(s) + H_2(g)$$

The chloride is easily hydrolysed.

Iron(II) oxide is obtained by heating iron(II) ethanedioate in the absence of air. A black powder, it is rapidly oxidized to iron(III) oxide. It is a basic oxide forming iron(II) salts with acids.

$$(COO)_2Fe(s) \longrightarrow FeO(s) + CO(g) + CO_2(g)$$

Iron(II) hydroxide is really hydrated iron(II) oxide. It is white when pure but is usually precipitated as a dark-green gelatinous precipitate which rapidly becomes brown as it is oxidized to hydrated iron(III) oxide.

Iron(III) compounds

1 **Iron(III) oxide**, Fe_2O_3 – jewellers' rouge. Occurs as haematite ore and is prepared by heating iron(II) sulphate crystals.

$$2FeSO_4.7H_2O(s) \longrightarrow Fe_2O_3(s) + SO_2(g) + SO_3(g) + 7H_2O(l)$$

It is amphoteric, resembling Al_2O_3 and Cr_2O_3. It forms iron(III) salts with acids and ferrates with alkalis, e.g.

$$Fe_2O_3(s) + 2OH^-(aq) \rightleftharpoons 2FeO_2^-(aq) + H_2O(l)$$

It is reduced to iron by carbon.

2 **Iron(II) diiron(III)** oxide, Fe_3O_4, occurs as magnetite, and lode-stone. It is obtained by heating iron in a current of steam or oxygen. It is a **compound oxide** behaving like iron(II) and iron(III) oxides.

$$3Fe(s) + 4H_2O(g) \rightleftharpoons Fe_3O_4(s) + 4H_2(g)$$

$$3Fe(s) + 2O_2(g) \longrightarrow Fe_3O_4(s)$$

$$Fe_3O_4(s) + 8HCl(conc.) \longrightarrow 2FeCl_3(aq) + FeCl_2(aq) + 4H_2O(l)$$

3 **Iron(III) hydroxide**, $Fe(OH)_3$, is precipitated as a reddish-brown gelatinous solid by adding aqueous alkali or ammonia to a solution of an iron(III) salt. It is feebly basic, forming iron(III) salts but the compounds are hydrolysed e.g.

$$Fe(OH)_3(s) + 3HCl(aq) \rightleftharpoons FeCl_3(aq) + 3H_2O(l)$$

4 **Iron(III) chloride**, $FeCl_3$, is obtained as a black sublimate by heating iron wire in dry chlorine.

$$2Fe(s) + 3Cl_2(g) \longrightarrow 2FeCl_3(s)$$

Up to 670 K, its formula is Fe_2Cl_6 but above this it dissociates to $FeCl_3$. It is soluble in ethoxyethane and its bonding is mainly covalent like that of $AlCl_3$.

Iron(II) $\rightleftharpoons$ Iron(III)

Iron(II) ions are always hydrated in solution as the **hexaaquairon(II) ion**, $[Fe(H_2O)_6]^{2+}$, which is pale green. It is easily oxidized to iron(III) but is slightly more stable in acidic solutions.

Iron(II) is oxidized to iron(III) by chlorine or bromine water, hydrogen peroxide solution, acidified potassium manganate(VII) or acidified potassium dichromate(VI) solution, e.g.

$$5Fe^{2+}(aq) + MnO_4^-(aq) + 8H^+(aq) \longrightarrow$$
$$5Fe^{3+}(aq) + Mn^{2+}(aq) + 4H_2O(l)$$

$$6Fe^{2+}(aq) + Cr_2O_7^{2-}(aq) + 14H^+(aq) \longrightarrow$$
$$6Fe^{3+}(aq) + 2Cr^{3+}(aq) + 7H_2O(l)$$

Iron(III) $\rightarrow$ iron(II)

Iron(III) ion is hydrated as the **pale purple** $[Fe(H_2O)_6]^{3+}$ but if the solution is not strongly acidic this rapidly hydrolyses forming aqua/hydroxo complexes, e.g. $[Fe(H_2O)_5OH]^{2+}$, $[Fe(H_2O)_4(OH)_2]^+$ and $[Fe(H_2O)_3(OH)_3]$ and finally $Fe_2O_3.xH_2O$.

Iron(III) is reduced to iron(II) by Zn/HCl or Zn/H_2SO_4, hydrogen sulphide, sulphur dioxide and potassium iodide solution, e.g.

$$2Fe^{3+}(aq) + S^{2-}(aq) \longrightarrow 2Fe^{2+}(aq) + S(s)$$
$$2Fe^{3+}(aq) + 2I^-(aq) \longrightarrow 2Fe^{2+}(aq) + I_2(\text{in KI soln.})$$

Rusting of iron occurs only in the presence of both oxygen and liquid water. It is accelerated by electrolytes and slowed down by alkalis. Completely homogeneous iron does not rust. In practice anodic areas and cathodic areas form on the surface of the iron.

Anodic: $\qquad Fe(s) \longrightarrow Fe^{2+}(aq) + 2e^-$

Cathodic: $\quad O_2(g) + 2H_2O(aq) + 4e^- \rightleftharpoons 4OH^-(aq)$

$$2H^+(aq) + 2e^- \longrightarrow H_2(g)$$

The reaction between iron(II) from anodic areas and hydroxide ions from cathodic areas form insoluble iron(II) hydroxide which is then rapidly oxidized to hydrated iron(III) oxide which is seen as rust. Corrosion is inhibited by giving the iron object a small negative potential or by coupling the iron with a more electropositive metal, usually zinc-galvanizing. The metal is protected by a thin adherent layer of zinc carbonate. Protection is maintained when scratched because zinc is anodic with respect to iron. Tin is preferred to zinc for food containers --

tinplate; but when scratched, iron coated with tin rusts faster because tin is less electropositive than iron.

Copper

Occurrence   Copper is found free and combined as the oxides Cu_2O and CuO, as a basic carbonate, malachite $CaCO_3.Cu(OH)_2$ and azurite $2CuCO_3.Cu(OH)_2$ and as its sulphide $CuFeS_2$.

Extraction – from copper pyrites $CuFeS_2$.

1   **The crushed ore is concentrated** by flotation (selective wetting). The powdered ore is agitated with alkaline water containing a frothing agent and copper-bearing particles collect in the froth.

2   **The ore is roasted** and iron is oxidized to iron(II) oxide leaving copper as copper(I) sulphide.

$$2CuFeS_2(s) + 4O_2(g) \longrightarrow Cu_2S(s) + 2FeO(s) + 3SO_2(g)$$

The sulphur dioxide is used to make sulphuric acid.

3   **The metal oxides are heated with sand** (silica) in a reverberatory furnace. Iron(II) silicate forms as a slag floating on top of a fused matter of mainly copper(I) sulphide and a little iron(III) sulphide. The slag is poured off.

$$FeO(l) + SiO_2(l) \longrightarrow FeSiO_3(l)$$

4   **The fused matter is run into a converter lined with magnesite,** $MgCO_3$. Silica is added and a blast of air blown through the mixture. Non-metals, e.g. sulphur, are oxidized and remaining iron forms a slag of silicate. Some of the copper(I) sulphide is oxidized to copper(I) oxide which reacts with the remaining copper(I) to form molten copper which is run off.

$$2Cu_2S(l) + 3O_2(g) \longrightarrow 2Cu_2O(l) + 2SO_2(g)$$
$$Cu_2S(l) + 2Cu_2O(l) \longrightarrow 6Cu(l) + SO_2(g)$$

As the molten copper curls, trapped sulphur dioxide escapes giving the metal a blistered appearance – **blister copper**.

5   **The copper is refined electrolytically.** The crude metal is made the anode and the cathode is a sheet of pure copper. The electrolyte is a solution of copper(II) sulphate with 5% sulphuric acid.

At the anode:

$$Cu(s) \longrightarrow Cu^{2+}(aq) + 2e^-$$

Copper passes into solution.

At the cathode:

$$Cu^{2+}(aq) + 2e^- \longrightarrow Cu(s)$$

Copper is deposited from the solution.

Copper may also be extracted by bacterial action.

Uses of copper
1 Copper is a very good conductor of heat and electricity – used in electrical circuits.
2 Used in alloys, e.g. bronze (originally Cu/Sn now coinage bronze contains Cu 95%, Sn 4%, Zn 1%), brass (Cu/Zn) and silver coinage (once contained silver) is now Cu/Ni.

Physical properties
Reddish-brown metal, very good conductor of heat and electricity, malleable and ductile.

Chemical properties
1 Copper is less electropositive than hydrogen and will not displace hydrogen from acids. **Hot concentrated sulphuric acid** attacks the metal readily.

$$Cu(s) + 2H_2SO_4(l) \longrightarrow CuSO_4(aq) + 2H_2O(l) + SO_2(g)$$

The products with **nitric acid** depend on the concentration of the acid. **With cold conc.** HNO_3,

$$Cu(s) + 4HNO_3(conc.) \longrightarrow$$
$$Cu(NO_3)_2(aq) + 2H_2O(l) + 2NO_2(g)$$

With cold moderately conc. HNO_3,

$$3Cu(s) + 8HNO_3(aq) \longrightarrow$$
$$3Cu(NO_3)_2(aq) + 4H_2O(l) + 2NO(g)$$

2 Copper becomes coated with a green patina on exposure to the atmosphere – mainly basic copper(II) sulphate, $CuSO_4.3Cu(OH)_2$.
3 If heated in dry air, copper oxidizes to Cu_2O and CuO.
4 Copper reacts readily on heating with halogens and sulphur.
5 Most of the chemistry of copper is concerned with the Cu^{2+} ion.

Compounds of copper

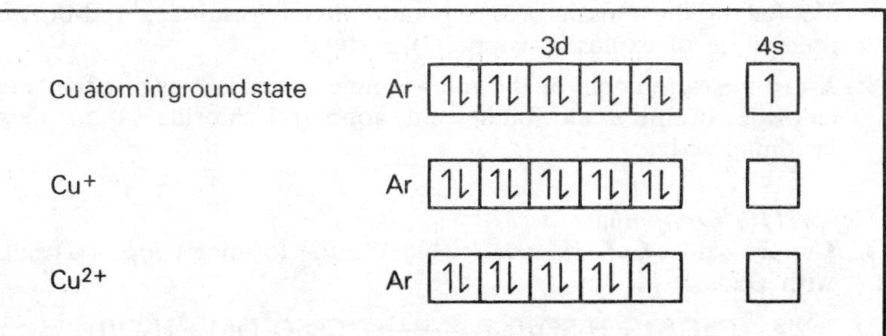

The copper(I) ion, Cu^+, has a completed d-subshell and no unpaired electrons. It is diamagnetic and its compounds are colourless (Cu_2O is red). The copper(II) ion, Cu^{2+}, has one unpaired electron, its compounds are paramagnetic and coloured.

Disproportionation of the Cu^+ ion

Cu^+ is hardly ever found in solution because:

$$2Cu^+(aq) \rightleftharpoons Cu^{2+}(aq) + Cu(s)$$

Copper(I) compounds

1   Sulphate, nitrate and carbonate do not exist in aqueous solution, any attempt to make them results in disproportionation, e.g.

$$Cu_2O(s) + H_2SO_4(aq) \longrightarrow CuSO_4(aq) + H_2O(l) + Cu(s)$$

2   **Copper(I) oxide**, Cu_2O Copper(II) sulphate solution is added to a solution of sodium hydroxide and sodium potassium tartrate (Rochelle salt) – Fehling's solution. When a glucose solution is added, a reddish-brown precipitate of copper(I) oxide is formed. Glucose acts as the reducing agent. It is basic but the copper(I) salts may disproportionate.

$$Cu_2O(s) + 2HCl(conc.) \longrightarrow 2CuCl(aq) + H_2O(l)$$

$$CuCl(aq) + Cl^-(aq) \longrightarrow [CuCl_2]^-(aq)$$

Other complexes, e.g $[CuCl_3]^{2-}$ may form.

3   **Copper(I) chloride** is precipitated when sulphur dioxide is passed through a solution containing equimolecular quantities of copper(II) sulphate and sodium chloride.

$$2Cu^{2+}(aq) + 2Cl^-(aq) + SO_2(g) + 2H_2O(l) \longrightarrow$$
$$2CuCl(s) + 4H^+(aq) + SO_4^{2-}(aq)$$

White when pure, but rapidly becomes green due to oxidation to basic copper(II) chloride. Insoluble in water but dissolves in concentrated hydrochloric acid and in ammonia, e.g. $[CuCl_2]^-(aq)$ and $[Cu(NH_3)_2]^+(aq)$. The ammoniacal solution absorbs carbon monoxide (as in the Bosch process) and ethyne forming a red-brown precipitate of explosive copper(I) acetylide.

4   X-ray measurements of the inter-atomic distances suggest that the chloride, bromide and iodide, e.g. copper(I) chloride vapour may be dimerized.

Copper(II) Compounds

1   **Copper oxide**, CuO, is a basic, black oxide forming copper(II) salts with acids, e.g.

$$CuO(s) + H_2SO_4(aq) \longrightarrow CuSO_4(aq) + H_2O(l)$$

It is readily reduced to reddish-brown powdery copper when heated in hydrogen or with carbon.

2 **Copper(II) hydroxide**, $Cu(OH)_2$, is precipitated as a blue gelatinous solid by the action of sodium or potassium hydroxide solution on copper sulphate solution. It dissolves in acids to form copper(II) salts and is weakly dissolving in excess concentrated alkali solutions to form the blue tetrahydraoxocuprate(II) ion, $[Cu(OH_4]^{2-}(aq)$.

3 The precipitate obtained by adding ammonia solution to copper(II) sulphate solution is pale blue and with excess ammonia forms a deep blue solution containing the diaquatetraammine copper(II) ion – Schweitser's reagent used in the manufacture of artificial fibres.

$$Cu(OH)_2(s) + 4NH_3(aq) + 2H_2O(l) \longrightarrow$$
$$[Cu(NH_3)_4(H_2O)_2]^{2+} + 2OH^-(aq)$$

4 **Copper(II) chloride**, $CuCl_2$, is obtained by heating copper in dry chlorine. On heating it dissociates to copper(I) chloride and chlorine. The chloride is soluble in water forming a blue solution, the colour of the solution depends on the concentration. More concentrated solutions are yellowish-green and on addition of concentrated hydrochloric acid the solution becomes brown. In the presence of excess chloride ions, forms the yellow-brown complex ion $[CuCl_4]^{2-}$. On dilution, the ion breaks up and blue hydrated $[Cu(H_2O)_6]^{2+}$ is formed.

Copper carbonate No true carbonates formed but basic carbonates, e.g. green $CuCO_3.Cu(OH)_2$ and blue $2CuCO_3.Cu(OH)_2$ exist.

Copper(II) compounds tend to be electrovalent, containing Cu^{2+} ions which in solution are usually hydrated, $[Cu(H_2O)_6]^{2+}$. The copper(II) salts are usually green in concentrated solutions and blue in dilute solution, CuO is black.

Zinc

Occurrence Found as zinc blende, ZnS and calamine, $ZnCO_3$.

Extraction
The sulphide is usually found mixed with galena, PbS.
1 **The crushed ore is concentrated** by flotation.
2 **The concentrated ore is roasted** in air when the calamine decomposes.

$$ZnCO_3(s) \longrightarrow ZnO(s) + CO_2(g)$$

The blende is oxidized by oxygen from the air.

$$2ZnS(s) + 3O_2(g) \longrightarrow 2ZnO(s) + 2SO_2(g)$$

The zinc oxide is left as a sintered solid and the sulphur dioxide is used in the manufacture of sulphuric acid.

3　The zinc oxide is reduced. It is mixed with coke and limestone and heated in a blast furnace. The zinc distils off and the vapour is condensed.

$$C(s) + O_2(g) \longrightarrow CO_2(g)$$

$$CO_2(g) + C(s) \longrightarrow 2CO(g)$$

$$ZnO(s) + CO(g) \longrightarrow Zn(g) + CO_2(g)$$

$$ZnO(s) + C(s) \longrightarrow Zn(g) + CO(g)$$

Uses

1　The metal is used in galvanizing – the iron article is dipped in powdered zinc under pressure, or the zinc is sprayed on.
2　In alloys, brass Zn/Cu; die-casting alloy, Zn/little Al/Cu/very little Mn; German silver, Zn/Cu/Ni.
3　In dry electric batteries, Zn is the cathode.
4　Used in the extraction of metals, e.g. less electropositive silver and gold are displaced.

Physical properties

Grey-white metal, brittle at room temperature. On heating becomes malleable and then malleable and ductile at higher temperatures.

Chemical properties

1　All 3d-orbitals are full in the zinc atom and divalent ion. Zinc does not behave as a typical transition element except in the formation of complexes.
2　Forms a thin, coherent layer of basic carbonate on exposure to air, which protects the metal from further action. Not affected by dry air. On heating catches fire with a green flame forming zinc oxide.
3　Zinc has little effect on water but decomposes steam at red-heat forming zinc oxide.

$$Zn(s) + H_2O(g) \longrightarrow ZnO(s) + H_2(g)$$

4　Concentrated sulphuric acid has little effect on zinc in the cold and on heating gives a variety of products including sulphur dioxide, hydrogen sulphide and sulphur.
5　Nitric acid forms a variety of compounds depending on the concentration and temperature of the acid – zinc nitrate with nitrogen dioxide, nitrogen oxide and ammonium compounds.
6　Powdered zinc react with **hot aqueous alkali** to form zincates and hydrogen.

$$Zn(s) + 2OH^-(aq) + 4H_2O(l) \longrightarrow$$

$$[Zn(OH)_4(H_2O)_2]^{2-}(aq) + H_2(g)$$

7 Combines when heated with **chlorine**, **sulphur** and **nitrogen**.
8 Strongly electropositive and precipitates less electropositive metals from solutions of their salts.

Compounds of zinc

Zinc oxide, ZnO   Yellow when hot, white when cold, reduced by carbon at red-heat to zinc, not reduced by hydrogen. Used as a white pigment (does not blacken in the presence of hydrogen sulphide). Used as a filler in rubber, in glazing porcelain and in zinc ointment.

Zinc hydroxide, $Zn(OH)_2$ Precipitated as a white gelatinous compound by caustic alkali, soluble in excess alkali.

$$Zn^{2+}(aq) + 2OH^-(aq) \longrightarrow Zn(OH)_2(s)$$

Amphoteric, forming salts with acids and the zincate ion, a hydroxo-complex with alkalis.

$$Zn(OH)_2(s) + 2OH^-(aq) + 2H_2O(l) \longrightarrow [Zn(OH)_4(H_2O)_2]^{2-}(aq)$$

The zinc hydroxide precipitated forms a zincate ion, tetrahydraoxozincate(II) with alkalis.

$$Zn(OH)_2(s) + 2OH^-(aq) + 2H_2O(l) \longrightarrow [Zn(OH)_4(H_2O)_2]^{2-}(aq)$$

The hydroxide precipitated by dilute ammonia from a zinc salt solution dissolves in excess ammonia due to the formation of a complex tetraammine ion, $[Zn(NH_3)_4]^{2+}$. Zinc hydroxide is not precipitated by ammonia when the solution is buffered by ammonium chloride.

Zinc carbonate is formed by the action of sodium hydrogencarbonate on the solution of a zinc salt. Sodium carbonate forms a basic carbonate.

$$3Zn^{2+}(aq) + CO_3^{2-}(aq) + 4OH^-(aq) + 2H_2O(l) \longrightarrow$$
$$ZnCO_3.2Zn(OH)_2.2H_2O(s)$$

Zinc chloride has a low melting point and is soluble in non-aqueous solvents – covalent character. It is hydrolysed in water.

Practice questions

1 The transitional elements arise from the gradual filling of penultimate d-orbitals. Write brief notes on the characteristics of the elements due to this electronic configuration. Consider: atomic radii, formation of alloys, densities, melting and boiling points, valency and bonding.

2 Suggest why the transitional elements (a) show paramagnetism, (b) form coloured ions.

3 Outline the extraction of iron.

4 Outline the process of basic oxygen steel-making.

5 What is (a) cast iron, (b) wrought iron, (c) steel?

6 How does iron react with (a) sulphur, (b) chlorine, (c) hydrogen chloride?

7 Write notes on (a) oxides of iron, (b) hydroxides of iron, (c) chlorides of iron.

8 Consider $Fe^{2+} \leftrightarrow Fe^{3+} + e^-$. Outline the conditions which determine this reaction.

9 Outline the conditions that cause rusting. How is it prevented? What is a sacrificial metal?

10 How is copper extracted from copper pyrites?

11 Suggest why copper(I) compounds are colourless while copper(II) compounds are usually coloured.

12 Write notes on (a) the oxides, (b) the hydroxides, (c) chlorides of copper.

13 How is zinc extracted?

14 Zinc does not have the general characteristics of a transitional element. Suggest why this is so.

15 How does zinc oxide react with (a) dilute hydrochloric acid, (b) sodium hydroxide solution, (c) ammonia?

13 OTHER d-BLOCK ELEMENTS

Ag		$4d^{10}$ $5s^1$
Cd	krypton	$4d^{10}$ $5s^2$
Hg	core	$4d^{10}$ $4f^{14}$ $5s^2$ $5p^6$ $5d^{10}$ $6s^2$

Silver

Occurrence Found uncombined, as silver glance, AgS, and horn silver, $AgCl$.

Extraction

Cyanide process Finely powdered ore is treated with sodium cyanide solution forming the dicyanoargentate(I) ion which is soluble.

$$Ag^+(aq) + 2CN^-(aq) \rightleftharpoons [Ag(CN)_2]^-(aq)$$

The liquid is decanted and zinc shavings added to displace the silver. The silver is refined by fusion with potassium nitrate, or electrolytically.

$$2[Ag(CN)_2]^-(aq) + Zn(s) \longrightarrow Zn^{2+}(aq) + 4CN^-(aq) + 2Ag(s)$$

Uses
1 In tableware and ornaments, jewellery.
2 In electroplating, the article to be plated being the cathode and pure silver being the anode.
3 In photography.

Physical properties
Soft white metal, very malleable and ductile, best known thermal and electrical conductor (but very costly).

Chemical properties
1 Only the oxidation state +1 is important, except for AgF_2.
2 Chemically resistant but is attacked by oxidizing acids.
3 Does not combine directly with oxygen but just above its melting point it absorbs about twenty times its volume of oxygen, which is released on cooling – spitting.
4 Becomes blackened with a layer of silver(I) sulphide on exposure to air.
5 Not affected by alkalis.
Silver(I) oxide, Ag_2O, is precipitated by the addition of alkali to a solution of silver(I) nitrate.

$$2Ag^+(aq) + 2OH^-(aq) \longrightarrow Ag_2O(s) + H_2O(l)$$

The hydroxide is not formed. Aqueous ammonia dissolves the precipitate forming the diammine-silver(I) complex.

$$Ag^+(aq) + 2NH_3(aq) \rightleftharpoons [Ag(NH_3)_2]^+(aq)$$

Cadmium

Occurrence   As sulphide with zinc blende. It is found in the zinc in the distillate obtained when zinc blende is reduced. The cadmium, which has a lower boiling point than zinc, is separated from the zinc by fractional distillation.

Uses   In plating steel, in alloys and as a neutron-absorber in nuclear reactors.

Properties   Soft, white metal, very poisonous. Forms salts with dilute acids and is not affected by alkalis.

Mercury

Occurrence   As cinnabar, HgS.

Extraction   The crushed ore is concentrated by flotation and then roasted in air with lime or iron to remove sulphur. Mercury vapour distils off and is condensed.

$$HgS(s) + O_2(g) \longrightarrow Hg(g) + SO_2(g)$$

The metal is purified by repeating distillation under reduced pressure.

Physical properties
Silvery liquid with a bluish tinge. The vapour and soluble compounds are very poisonous.

Chemical properties
1   When heated to just below its melting point, red mercury(II) oxide, HgO, is formed. This decomposes at higher temperatures.
2   Attacked by oxidizing acids, e.g.

$$2Hg(l) + 2H_2SO_4(l) \longrightarrow Hg_2SO_4(s) + 2H_2O(l) + SO_2(g)$$

3   No reaction with hydrochloric acid or alkalis.
4   Combines directly on heating with chlorine.

$$Hg(l) + Cl_2(g) \longrightarrow HgCl_2(s)$$

6   Forms green mercury(I) iodide when rubbed with iodine.

$$2Hg(l) + I_2(s) \longrightarrow Hg_2I_2(s)$$

7   Amalgamates with many metals, Na/Hg is used in the manufacture of sodium hydroxide.

Uses   As the cathode in electrolytic processes, in the production of vermilion HgS, mercury fulminate $Hg(NO)_2$ (detonator), in dental fillings and in electrical switches.

Compounds of mercury

Hg^+ does not exist and mercury(I) compounds contain Hg_2^{2+} containing a metal-metal covalent bond, $^+Hg{-}Hg^+$. It tends to disproportionate.

$$Hg_2^{2+}(aq) \longrightarrow Hg(l) + Hg^{2+}(aq)$$

Mercury(I) salts are stabilized by adding a little mercury metal. A mixture of mercury and mercury(II) chloride, in the absence of water, yields a sublimate of mercury(I) chloride.

$$Hg(l) + HgCl_2(s) \longrightarrow Hg_2Cl_2(g)$$

A solution of mercury(I) ions is obtained by the action of dilute nitric acid on excess mercury.

$$6Hg(l) + 8H^+(aq) + 2NO_3^- \longrightarrow 3Hg_2^{2+} + 2NO(g) + 4H_2O(l)$$

Hot concentrated nitric acid forms mercury(II) nitrate.

Mercury(II) oxide, HgO, is obtained when an alkali is added to a solution containing Hg^{2+} or when mercury is gently heated in air. The yellow form becomes red on warming. It is a basic oxide.

Mercury(II) chloride, $HgCl_2$, is a white, very poisonous solid formed when mercury(II) oxide dissolves in hydrochloric acid. It sublimes on heating mercury in dry chlorine or a mixture of mercury(II) sulphate and sodium chloride. Covalent, low conductivity and soluble in ethoxy-ethane. Addition of excess chloride ion forms the complexes $[HgCl_3]^-$ and $[HgCl_4]^{2-}$.

Mercury(II) iodide is precipitated when potassium iodide is slowly added to mercury(II) chloride in solution. It is yellow at first but turns red. Excess potassium iodide dissolves the precipitate to form the complex tetraiodomercury(II) ion.

$$HgCl_2(aq) + 2KI(aq) \longrightarrow HgI_2(s) + 2KCl(aq)$$

$$HgI_2(s) + 2I^-(aq) \rightleftharpoons [HgI_4]^{2-}(aq)$$

When the solution is made alkaline in potassium hydroxide, the solution is Nessler's reagent – used to test for NH_4^+.

Practice questions

1 Outline the cyanide process for the extraction of silver.

2 Give the reaction(s) between silver nitrate and   (a) sodium hydroxide solution   (b) aqueous ammonia.

3 How is mercury extracted?

4 Hg^+ does not exist. Describe the mercury(I) ion in, for example, mercury(I) chloride.

5 Write notes on   (a) the oxide,   (b) the chloride and   (c) the iodide of mercury(II).

14   COMPLEXES AND DOUBLE SALTS

Addition or molecular compounds are formed by the stoichiometric combination of apparently saturated substances which can exist independently forming **double salts** or **complexes**.

A **double salt** exists only in the solid state and breaks up when dissolved in water, e.g. potash alum or alum. $KAl(SO_4)_2.12H_2O$.

A **complex** retains its identity in solution. It consists of a metal ion or atom surrounded by a number of oppositely charged ions or neutral molecules which can donate lone pairs of electrons to the vacant orbitals of the central ion or atom. The ion or molecule donating the electrons

is called a *ligand* as in, e.g. potassium hexacyanoferrate(III), $K_3[Fe(CN)_6]$, in which the iron(III) is surrounded by six CN^-.
Some double salts contain complex ions, e.g. $KAl(SO_4)_2.12H_2O$ is more accurately written $K^+[Al(H_2O)_6]^{3+}(SO_4)_2^{2-}6H_2O$ with $[Al(H_2O)_6]^{3+}$ existing when the compound is added to water. If the stability of the complex ion is low it may break up, e.g. when potash alum is dissolved in water, some aluminium ions are formed.
The presence of a complex is detected by measuring colligative properties which depend on the number of particles present – if a complex is present, the number of particles is fewer than expected.

Bonding and structure
1   The central atom or ion must be able to accept the lone pair or pairs of electrons.
2   Transition elements form complexes because they contain incomplete shells into which electrons can be donated. For the elements scandium to zinc, there are only small differences between the energy levels 3d, 4s and 4p so that 3d-electrons can be excited and used in bonding. The orbitals are hybridized so that the resulting bonds are all the same.

The coordination number of the central metal or ion is the total number of lone pairs that it can accept from the ligands.

Coordination number and shape
Coordination number 2, complex is linear, e.g. $[Ag(CN)_2]^-$; coordination number 4, complex is square planar, e.g. $[Ni(CN)_4]^{2-}$ or tetrahedral, e.g. $[Ni(CO)_4]$; coordination number 6, complex octahedral, e.g. $[Cr(H_2O)_6]^{3+}$.

Square planar complex dsp^2 hybridization

	3d	4s	4p
Ni atom in ground state	Ar [↑↓][↑↓][↑↓][↑][↑]	[↑↓]	[ ][ ][ ]
Ni^{2+}	Ar [↑↓][↑↓][↑↓][↑][↑]	[ ]	[ ][ ][ ]
Ni^{2+} in $[Ni(CN)_4]^{2-}$	Ar [↑↓][↑↓][↑↓][↑↓][×]	[×]	[×][×][ ]

dsp^2 hybridization

X represents a lone pair of electrons donated by CN^-

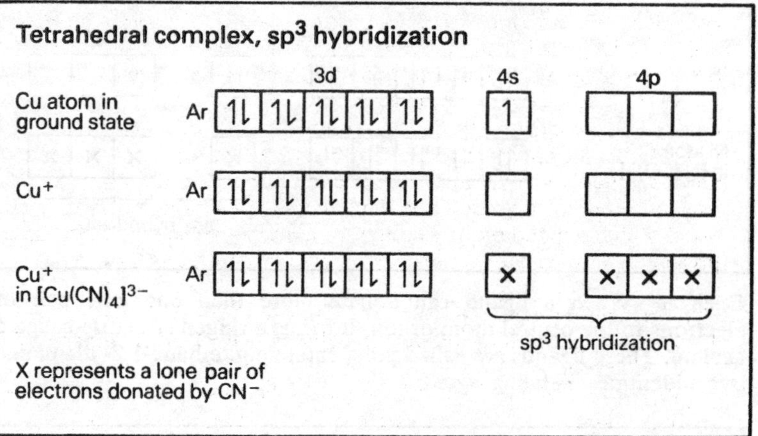

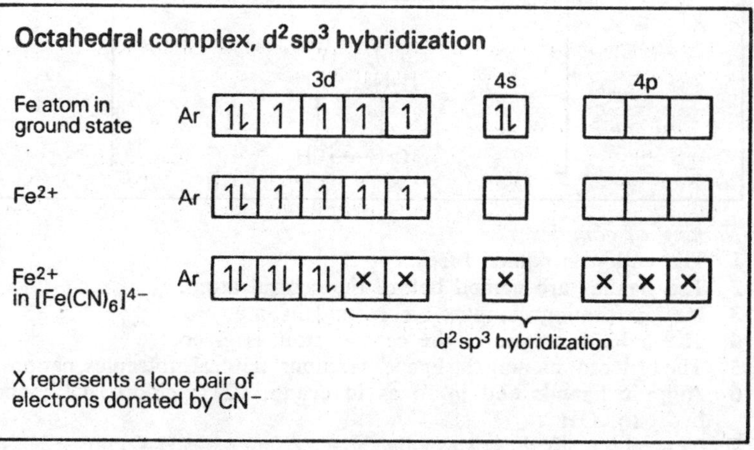

N.B. Fe^{2+} contains 4 unpaired electrons and is paramagnetic. The iron $[Fe(CN)_6]^{4-}$ is diamagnetic and therefore has no unpaired electrons. **Metal carbonyls** are formed by the bonding of neutral carbon monoxide with a central metal atom, e.g. $[Ni(CO)_4]$.

There is also some delocalization involving the carbon monoxide molecules.

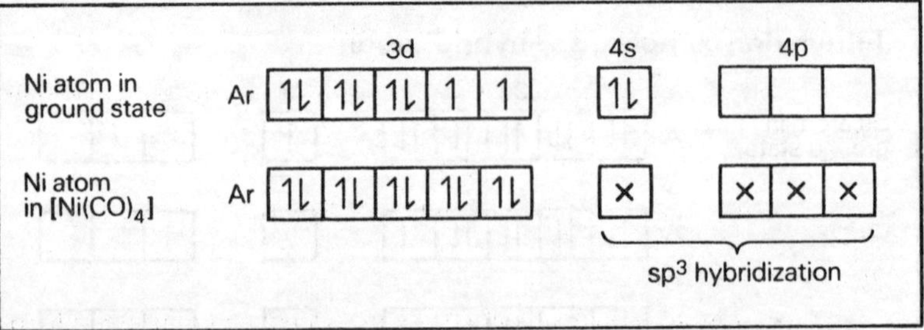

Chelates   When a ligand can donate more than one lone pair of electrons to the central atom or ion, it forms a ringed structure called a chelate. These ligands are called polydentate e.g. ethane-1,2- diammine is a bidentate chelating agent:

Naming of complexes
1   The cation is named first.
2   The ligands are named before the central atom.
3   Metals forming complex anions end in -ate.
4   The oxidation state of the central atom is given.
5   The order for naming the ligands is anions, neutral molecules, cations.
6   Anionic ligands end in -o as in cyano, CN^-; chloro, Cl^-; and hydroxo, OH^-.
7   Neutral ligands do not usually show a change but H_2O, aqua; NH_3, ammine; and CO, carbonyl.
8   Positive ligands end in -ium, e.g. NH_2—NH_3^+, hydrozonium.
9   If there is already a number in the ligand, then the ligand is placed in a bracket and the prefix bis, tris etc. is used instead of di and bi.

Isomerism
This arises from:
1   The same ions occupying different places, e.g. dichlorotetraammine-platinum(II) bromide $[Pt(NH_3)_4Cl_2]Br_2$ and dibromotetraammine-platinum(II) chloride, $[Pt(NH_3)_4Br_2]Cl_2$.

2 Hydration, e.g. $[Cr(H_2O)_6]Cl_3$ (violet), $[Cr(H_2O)_5Cl]Cl_2.H_2O$ (green), $[Cr(H_2O)_4Cl_2]Cl.2H_2O$ (green).

3 The presence of a ligand which has two atoms that can donate a lone pair, e.g. NO_2^-.

4 The interchange of ligands, e.g.

$$[Cu(NH_3)_6]^{2+}[PtCl_4]^{2-} \quad \text{and} \quad [Pt(NH_3)_4]^{2+}[CuCl_4]^{2-}$$

Practice questions

1 What is a complex? Explain why the transitional elements are able to form complexes.

2 Describe how you would obtain crystals of a named double salt.

3 Write brief notes on   (a) coordination number,   (b) ligands (c) chelation.

4 Name the following complexes:
(a) $[Cr(H_2O)_6]CCl_3$
(b) $Ni(CO)_4$
(c) $K_2[PtCl_4]$
(d) $[Pt(NH_3)_4]Cl_2$
(e) $[Co(NH_3)_5NO_2]Cl_2$

15   ANSWERS TO PRACTICE QUESTIONS

Atomic Structure

8 (a) (V), (VI) and (VII); (I) and (III)
(b) (II) and (III); (IV) and (V); (VII) and (VIII); (IX) and (X)
(c) (II) (IV) (VI); (VIII), (IX) and (X)

9 Cu^+ $2s^2$ $2p^6$ $3s^2$ $3d^{10}$ $4s^1$ paramagnetic
Cu^{2+} $2s^2$ $2p^2$ $3s^2$ $3d^9$ diamagnetic

10 $_{17}^{35}Cl : _{17}^{37}Cl = 3 : 1$ paramagnetic

s-BLOCK ELEMENTS: The Alkali Metals of Group I

1 $_3Li$ $1s^2$ $2s^1$
$_{11}Na$ $1s^2$ $2s^2$ $2p^6$ $3s^1$
$_{19}K$ $1s^2$ $2s^2$ $2p^6$ $3s^2$ $3p^6$ $4s^1$
$_{37}Rb$ $1s^2$ $2s^2$ $2p^6$ $3s^2$ $3p^6$ $3d^{10}$ $4s^2$ $4p^6$ $5s^1$

s-BLOCK ELEMENTS: The Alkaline Earth Metals of Group II

1 $_4Be$ $1s^2$ $2s^2$
$_{20}Ca$ $1s^2$ $2s^2$ $2p^6$ $3s^2$ $3p^6$ $4s^2$
$_{20}Ca^{2+}$ $1s^2$ $2s^2$ $2p^6$ $3s^2$ $3p^6$
$_{56}Ba$ $1s^2$ $2s^2$ $2p^6$ $3s^2$ $3p^6$ $3d^{10}$ $4s^2$ $4p^6$ $4d^{10}$ $5s^2$ $5p^6$ $6s^2$
$_{56}Ba^{2+}$ $1s^2$ $2s^2$ $2p^6$ $3s^2$ $3p^6$ $3d^{10}$ $4s^2$ $4p^6$ $4d^{10}$ $5s^2$ $5p^6$ $6s^2$

p-BLOCK ELEMENTS: Halogens of Group VII

4 (a) and (c) will not occur.

Complexes

4 (a) hexaaquochromium(III) chloride
(b) tetracarbonyl nickel
(c) potassium tetrachloroplatinate(II)
(d) tetraammineplatinum(II) chloride
(e) nitropentaamminecobalt(III) chloride

Periodic Table chart. Labels: s-block, Reactive metals (Groups I, II); Transition metals; d-block; f-block; p-block; Non-metals; Poor metals; Noble gases.

Period	Group I	Group II	d-block / transition										Group III	Group IV	Group V	Group VI	Group VII	Group O
Period 1									H 1									He 2
Period 2	Li 3	Be 4											B 5	C 6	N 7	O 8	F 9	Ne 10
Period 3	Na 11	Mg 12											Al 13	Si 14	P 15	S 16	Cl 17	Ar 18
Period 4	K 19	Ca 20	Sc 21	Ti 22	V 23	Cr 24	Mn 25	Fe 26	Co 27	Ni 28	Cu 29	Zn 30	Ga 31	Ge 32	As 33	Se 34	Br 35	Kr 36
Period 5	Rb 37	Sr 38	Y 39	Zr 40	Nb 41	Mo 42	Tc 43	Ru 44	Rh 45	Pd 46	Ag 47	Cd 48	In 49	Sn 50	Sb 51	Te 52	I 53	Xe 54
Period 6	Cs 55	Ba 56	La 57	Hf 72	Ta 73	W 74	Re 75	Os 76	Ir 77	Pt 78	Au 79	Hg 80	Tl 81	Pb 82	Bi 83	Po 84	At 85	Rn 86
Period 7	Fr 87	Ra 88	Ac 89	Ku 104	Ha 105													

f-block:

Ce 58	Pr 59	Nd 60	Pm 61	Sm 62	Eu 63	Gd 64	Tb 65	Dy 66	Ho 67	Er 68	Tm 69	Yb 70	Lu 71
Th 90	Pa 91	U 92	Np 93	Pu 94	Am 95	Cm 96	Bk 97	Cf 98	Es 99	Fm 100	Md 101	No 102	Lr 103

Periodic Table